Fishing Lessons

FISHING LESSONS
ARTISANAL FISHERIES AND
THE FUTURE OF OUR OCEANS

Kevin M. Bailey

The University of Chicago Press
Chicago and London

The University of Chicago Press, Chicago 60637
The University of Chicago Press, Ltd., London
© 2018 by Kevin M. Bailey
All rights reserved. No part of this book may be used or reproduced in any manner whatsoever without written permission, except in the case of brief quotations in critical articles and reviews. For more information, contact the University of Chicago Press, 1427 E. 60th St., Chicago, IL 60637.
Published 2018
Printed in the United States of America

27 26 25 24 23 22 21 20 19 18   1 2 3 4 5

ISBN-13: 978-0-226-30745-9 (cloth)
ISBN-13: 978-0-226-30759-6 (e-book)
DOI: https://doi.org/10.7208/chicago/9780226307596.001.0001

Library of Congress Cataloging-in-Publication Data

Names: Bailey, Kevin McLean, author.
Title: Fishing lessons : artisanal fisheries and the future of our oceans / Kevin M. Bailey.
Description: Chicago ; London : The University of Chicago Press, 2018. | Includes bibliographical references and index.
Identifiers: LCCN 2017039497 | ISBN 9780226307459 (cloth : alk. paper) | ISBN 9780226307596 (e-book)
Subjects: LCSH: Small-scale fisheries.
Classification: LCC SH329.S53 B35 2018 | DDC 338.7/6392—dc23
LC record available at https://lccn.loc.gov/2017039497

♾ This paper meets the requirements of ANSI/NISO Z39.48-1992 (Permanence of Paper).

Contents

Introduction 1

Fouled Fish

1. *The Giovanni Padre: The Sun Sets on Small-Scale Fisheries in the Gulf of Naples* 19
2. *The King Is Dead: The Collapse and Resurrection of Vosso Salmon* 35
3. *Ode to the Sea: Chile's Troubled Fisheries* 49

Loss and Recovery of Indigenous Fisheries

4. *The First Fish: The Coast Salish Salmon Fishery* 71
5. *Northern Lights: The Sea Sámi Fishery in Norway* 93

Return to Artisanal

6. *A Clean and Green Fishery: Legoe Bay Reefnets* 117
7. *Crimson Tide: The Bay of Fundy Weir Fishery and a Conflict with Green Power* 129
8. *A Dying Fishery? Puget Sound Keta Salmon* 145

Striking a Balance in Aqua Farming

9. *Mother of Pearl: Ocean Farming Red Abalone in Monterey Bay* 161
10. *King of the Amazon: Culture and Harvest of Arapaima* 177
11. *Evolving Solutions* 193

Acknowledgments 209
Notes 211
References 223
Index 241

Introduction

I watched Tonino Calise mend his nets in the afternoon sun. His leathered fingers braided the twine without the need of thought, or interruption. With a cigarette dangling from his lips like James Dean, he squinted up at me as I scribbled in my notebook. He offered brusque answers to my questions, then turned back to his work.

An Italian friend had introduced me to Tonino and translated for me as we talked about the small-scale artisanal fisheries of Ischia, a small island in the Bay of Naples. I was in Ischia trying to shed my scientist's skin and learn about the human side of fisheries. During my thirty-five years as a biologist experimenting in the laboratory, working on research ships at sea, and then more and more sitting in front of a computer screen, I'd had little exposure to the lives of fishermen.

Fishing is an ancient way of life; its citizens now are graying and disappearing, washed over by the changing world around them. The life of a small-scale traditional fisherman is challenging. This is rough, hard work. Not many young people are attracted to it. While a life of fishing offers independence, it can be lonely. But for those who are drawn to the boats, the salt of the ocean is in their blood, and the rhythm of the sea is never far away.

What manner of person ventures out in his boat to roam the sea, face danger, and suffer hardships? If traditional fishermen disappear, what happens to their acquired knowledge of the sea, their know-how of the boat, their expertise in catching fish? These are skill sets that will be hard to replace. Are small fisheries like Tonino's worth saving, or are they relics of the past? Do we value more plentiful and cheaper fish, or is it important to preserve a way of life? Are so many traditional fishermen the source of the overfishing problem, or can they be part of the solution for the ocean's troubled fisheries?

Fishing from his small open boat in the late afternoon, Tonino casts a small gillnet called a *schietta* into the sea. He usually catches Mediterranean hake, monkfish, mackerel, the occasional swordfish or tuna, and some other small species. In the early morning he retrieves his net and chugs back to the harbor at daybreak to sell his fish in the local market.

Tonino told me that he catches fewer fish now than in years past. He's had to resort to using a larger net. He blames the big trawlers from Procida, a small island nearby, and from the mainland port of Pozzuoli for catching too many fish, thereby leaving few behind for the other fishermen.

The island of Ischia sits on the edge of a protected marine reserve in the Mediterranean. I asked Tonino if the reserve has helped him as a fisherman. He responded with a laugh, sweeping his hand through the air as if to shoo away a fly, and remarked that the marine reserve has had little impact on the availability of fish in the area. He said the regulations in the reserve are lax and often not enforced. Everyone fishes there. The refuge is mainly present to attract scuba-diving tourists. When I mentioned the governance of fisheries by catch limits and individual quotas, the management tools put forth by academic scientists and government administrators to prevent overfishing, he scoffed. "It won't work here. There are too many fishermen, and they don't keep records."

Later in the evening when darkness fell, the air was muggy and I was restless. I opened the windows of my small rented apartment overlooking the harbor. The village of Lacco Ameno has become a tourist hot spot for the yachting crowd. Shoreline development has accommodated them, while decimating the local fisheries by destroying habitat and creating pollution. In the night air, at least three bands competed for sound space. The dissonance continued until the wee hours. One band played Mexican love songs; another was, I think, a klezmer jazz band with clarinet, tambourine, and thundering bass

Figure 1. Fisherman Tonino Calise mending his nets in the harbor of Lacco Ameno on the island of Ischia, Italy. Photographer: K. Bailey.

guitar. Floating on top of it all was a sultry voice singing Italian pop songs. My thoughts turned over Tonino's dismissal of the management measures offered by fisheries biologists (like myself) and other well-meaning desk jockeys astride Aeron chairs, often ignorant of the reality on the waterfront. The weaving strains of music hinted of a metaphor for the discord among the fishermen, government administrators, and scientists.

Not long after I returned home from Italy, my friend from Ischia wrote me a note to say that Tonino didn't return with his catch one morning. They had found his boat adrift. He was fishing alone when the heart attack struck, and he died on the ocean, doing what he loved. My all-too-brief encounter with Tonino symbolized a fading human legacy.

Plight of Traditional and Small-Scale Fishermen

Interpretations of "traditional fisheries" range from traditional subsistence fisheries to harvesting operations where fishermen use only traditional gear or even unpowered vessels. Some definitions involve the size of boat used in the fishery. Sometimes the terms *artisanal fishery* and *small-scale traditional fishery* are used interchangeably. I think a modern and practical definition would incorporate the ownership, scale, method of harvesting, and quality of the product delivered. For this book, *I define a traditional fishery as small in scale, owner-/family-operated, delivering a craft product to the market.* Often the method of harvest is customary. But since virtually all fishing gear being used now was available in some form prior to the Industrial Revolution, most gear is included here. Vessel size is generally less than twenty meters in length.

There is a blurry line between artisanal and small-scale traditional fishermen.[1] Most agencies don't discriminate between them. But in my view, the term *artisanal* has evolved such that not all small-scale traditional fishermen are artisanal. Some fishermen, regardless of the scale of their operation, are not concerned about the ecosystem, conservation, or responsibility to consumers. I consider that the new artisanal fisherman takes pride in the quality of product he provides and values the health of the ecosystem. The term *artisanal* reflects a way of life, practicing respect and passion for the craft of fishing.[2] An artisanal fishery is thereby a subset of small-scale traditional fisheries, although the fishermen of both groups often face similar problems.

Small-scale traditional (including artisanal) fishermen are struggling for their survival because of the escalating pressures of coastal development, industrial fishing, restricted areas, and government programs to pre-assign the harvest, often granting fishing rights to a privileged few. Small-scale fishermen can't compete for the catch against large, powerful industrial vessels with their big nets, advanced technology, and economic efficiency. There is a shift away from independent fishermen who catch and sell their harvest, toward the direction of wage earners who work on floating factories.

Tonino's Italy exemplifies the plight of small-scale traditional fishing. Many countries like Italy have seen increasing fishing effort because of the growth of the industrial fishing sector, while experiencing declines in catches and a sense of loss of tradition. In the Italian fishing fleet of about twenty thousand vessels, 63 percent are small scale and traditional. Since the end of World War II, the number of boats has declined but the total tonnage and engine power has markedly increased. The remaining boats are getting bigger and more powerful. The small boats are being sacrificed. From 1990 to 2000 alone there was a 25 percent decline in the small-scale fleet.

Traditional fishermen are losing the competition with industrial fishing companies not only at sea, but in the marketplace as well. Because of lowered overhead and high efficiency, big fishing companies can sell their products for less than the small-scale fishermen. Efficient supply lines of corporate fisheries get their products to the global marketplace readily and cheaply. Recently I came across a report that Alaska pollock caught in the Bering Sea, frozen on board, and then shipped to China for processing and refreezing, is sold in Brazil—nearly a hemisphere away—for less than half of the price of locally produced tilapia.[3] A recent report from Senegal says that local consumers who depend on fish for protein cannot compete with industrial buyers who reduce the fish to a dry meal that is fed to chicken, pigs, and salmon in Europe.[4]

Why should we be concerned? Fisheries are an important resource in the global economy. Capture fisheries[5] and aquaculture produced 148 million tons of fish in 2010, worth $217 billion. In 2009, 16.6 percent of the world's human intake of animal protein was fish. By 2050, it is projected that nine billion people will live on Earth. The world's food supply will have to increase by 70 percent to feed them. Furthermore, in 2010 fisheries and aquaculture provided a livelihood for 54.8 million people involved in production. Ancillary activities like processing, packaging, marketing, transporting, and supplying materials to fisheries enterprises supported another 660 million to 820 million people.[6]

Traditional fisheries are a livelihood and way of life for millions of people. They are important to maintain for food security, to benefit local communities, and to sustain cultural identity.[7] Small-scale fisheries, as opposed to industrial fisheries, contribute about one-half to two-thirds of the global food-fish catch (the harvest used to feed people directly rather than as feed for other animals), and employ from 80 to 90 percent of the world's fishermen and fish workers. About half of these workers are women.[8]

In theory, artisanal and traditional small-scale fisheries should be less destructive to ocean health than large-scale and industrial fisheries. Traditional gear is usually fixed in place or confined in space and tends to be less destructive to the habitat than large trawls. Traditional gear is often more selective, resulting in less bycatch (species caught incidentally to the targeted species). Often the bycatch of traditional fisheries is utilized as food or released alive, instead of thrown back into the sea, damaged or dead. Because traditional fishermen are part of their local community, their catches support the infrastructure of the local businesses neighboring them.

There is another important role of traditional fishermen and their communities. More and more, people who make their living from the sea are involved in defending the seas against industrial overfishing, destructive practices of the mining and energy-extraction industries, and shoreline development. Indigenous peoples and traditional fishing communities are taking the lead to defend their resources, shorelines, and seas against depredation from marauding outsiders.

The Past and Present of Fisheries

Humans have fished the sea in an organized way for at least forty-two thousand years; the earliest-known evidence is based on fish bones in caves of East Timor.[9] Based on archaeological records from Europe, fishing with "gorges" — sticks sharpened at both ends and attached to a line in the middle like primitive hooks — dates back thirty thousand years. Evidence of man fishing predates history. But now this most primal aspect of our relationship with the ocean, obtaining food from the sea, is changing rapidly. It's important to know what is happening and why, what we are losing, and what we are gaining as a global community.

Indigenous fisheries were originally subsistence based for personal or community use. A common theme across indigenous cultures was "take no more than you need." After methods were developed to preserve meat, dried and salted fish became a common item that coastal groups traded with neighboring people to acquire other necessities. Much later, the arrival of colonial powers from more developed cultures signaled another change to indigenous coastal communities. The colonials were interested not only in their own subsistence, but in acquiring resources and wealth besides. The theme changed from "take no more than you need" to "take as much as you can."

A mechanical revolution in the late 1870s brought another major change to the nature of fisheries. The forces powering fishing vessels shifted from wind and oars to steam-powered engines, then to diesel and turbines. This change produced larger ships that could travel to more distant parts of the ocean, as well as catch, hold, and carry more fish. Shipyards built boats with iron hulls and later with steel.

During the Second World War, a technological push to defeat the Axis powers led to breakthroughs in acoustics, vessel positioning, lightweight synthetic materials, refrigeration, and propulsion. After these innovations migrated from the military to civilian industries, fishermen could find fish more easily, tow larger nets, and preserve the catch better. Now they could stay longer at sea, travel greater distances, and expand their markets.

As boats got bigger, the fishing communities and markets evolved to adapt to the new conditions. Solo fishermen gave way to harvesting teams that now included a crew of professional fishermen and a group of investors. Traders bought the fish and sent them to distant lands, establishing complicated networks of middlemen. Because vessel sizes continued to increase and technology improved, ships and gear became more expensive. Financial investments became greater, but so did the rewards. Fishing companies grew larger and more sophisticated, often incorporating with a board of directors to control the fishery from shore and in-house attorneys to protect their interests.

Fisheries development has now become so sophisticated that Wall Street has taken notice and private equity funds own the harvest rights of some fish stocks.[10] For many present-day fisheries (like Alaska pollock—the world's largest food fishery) floating factories and fleets of vessels controlled by large corporations take most of the harvest. The fish are headed and gutted belowdecks, often by immigrants from underdeveloped nations; the meat is flash-frozen, and then much of it is shipped to China for further processing. From there it is sent around the world or even back to the United States for consumption.

During an era of expanding fisheries, from 1950 to 1980, global fish catches quadrupled. Increasing at about 8 percent per year, fisheries growth far surpassed the human population growth of 2.5 percent per year. Marine capture fisheries were promoted as a way to feed the world. Governments subsidized building large industrial fishing vessels. However, the supply of fish in the ocean is finite, and the growth in fisheries wasn't to last; by 1990, the expansion was over. The harvests leveled off, growing only 1 percent per year until 2000. In recent years, catches even declined, resulting in too few fish for too many boats. The overcapacity of the fishing fleet wasn't due to an increase in the number of small traditional boats, but rather, from the larger ships of the industrial sector. Faced with a shortage of fish, the now-huge fishing capacity of large vessels squeezed into the nearshore-coastal areas of traditional fishermen.

The modern history of fishing is a classic case of resource exploitation: discovery, developing skill and technology, expansion, and finally limitation. The 2012 report of the United Nations Food and Agriculture Organization (FAO) on the state of world fisheries says, "The declining global marine catch over the last few years together with the increased percentage of overexploited fish stocks conveys a strong message that the state of world marine fisheries is worsening."[11] But as in most situations where money and resources collide, there is an opposing viewpoint that the state of world fisheries isn't as bad as depicted, and in fact, is steadily improving. Advocates believe that fish harvests can increase. This opinion is often supported by the fishing industry that has the most to gain.

Who Owns the Fish in the Sea?

The prospect of declining fish stocks—too many fishermen and not enough fish—has driven the most recent revolution in fisheries. Unlike what is happening in most other industries, the new revolution in fisheries is not a technology-driven change, but represents an ideological shift.

The bounty of the ocean has, until recently, been a publicly owned resource. To prevent overfishing—catching more fish than the population can sustain—governments controlled fishermen by measures such as placing restrictions on their gear, or where and when they were allowed to fish. Eventually a system of limited entry developed to control how many boats were fishing. Most recently, economists developed the idea of pre-assigning ownership of quotas, or catch shares. Now, not only the rights to fish but the "ownership" of a defined proportion of the available fish are being given to specific individuals and companies, usually for free. The ocean's wealth is being redistributed, and in the view of many, the small-scale fishermen are losing out.

How did we get to this state? Indigenous communities have always had forms of territorial rights, often involving shared, proprietary, or bartered rights for natural resources. About two thousand years ago the Romans pronounced that the marine realm, including its fish, was the common property of all—meaning all Roman citizens. Under the Roman plan, the sovereignty of nations, or of individuals, over the seas extended only to the high-tide mark.

As the Roman Empire weakened in the Middle Ages, the rulers of coastal lands pushed their sovereignty seaward. By the fifteenth and sixteenth centuries, Portugal and Spain claimed large portions of the oceans. In 1455 the Catholic Church gave Portugal exclusive control of the African coast. When Vasco da Gama sailed around the Cape of Good Hope in 1488, Portugal claimed most of the trade through the Indian Ocean. In 1493 Spain got the rights to the New World by papal declaration. Other countries began to grumble about the domination of Spain and Portugal. By the end of the sixteenth century, England, the Netherlands, and France were challenging the control of the oceans held by the Iberian countries. England claimed sovereignty over her own seas and the fish within, while arguing for the freedom of the rest of the oceans.

Portugal blocked the Dutch East India Company from opening trade routes around Africa to the spice-rich lands in the Indian Ocean. After the Dutch seized a Portuguese merchant ship on the basis that Portugal was limiting free access to the sea, they engaged a young attorney named Hugo Grotius to write his manifesto on the freedom of the sea, called *Mare Liberum*, in 1609. The concept of the freedom of the seas from the Roman era was reintroduced. Grotius argued that because the sea was free-flowing and limitless, like the air we breathe, no entity could own it or the fish within.

This principle became widely accepted; however, as time passed, the coastal boundaries of many nations were pushed out to three miles, said to be the distance a cannonball could fly in defense of territorial claims. Over the coming centuries, one country after another pushed their claims to include national ownership of resources within coastal territories, extending their boundaries farther out—to six miles, then to twelve miles, and finally to two hundred miles by unilateral action, culminating in the United Nations' "Law of the Sea." Beyond that, the seas were still free.

Once national sovereignty was established, governments were able to control the nature of their fisheries, including who could fish and how much they could harvest. Generally, fishing activity has been regulated by nation-states or local governments that issue licenses and permits. Increasingly, there has been the need to control the number of boats fishing.

Over the past three decades, a new tool for nations to regulate fisheries within their sovereign territory has become popular. The so-called catch share programs are seen by some as a way to transfer the rights of access to ocean resources from public to private entities. A catch share is a set percentage of the allowable harvest of a fish stock that is granted to an entity (person, vessel, cooperative, or community) prior to the harvest season. In one popular version of catch share programs, called *individual transferable fisheries quotas* (commonly known as IFQs or ITQs), the catch share can be traded, leased, sold, and inherited just like property. Usually the initial assignment of catch shares is based on prior fishing history of the entities involved. Sometimes the shares are sold or leased to new owners who have never even fished, and may never do so. Other programs have some restrictions on who can own and fish quotas.

The catch share programs have been championed by some powerful nongovernmental organizations like the Environmental Defense Fund (EDF), which orchestrates a well-funded public relations campaign. They receive financial support from foundations that often promote privatization of public resources, such as the Charles Koch Foundation, the David and Lucile Packard Foundation, and the Walton Family Foundation.[12] Catch shares can be publicly owned and auctioned to fishermen, not unlike timber or mining leases on public land; but based on social ideologies, these foundations have pushed for private ownership rights, similar to property rights. Catch share programs in various forms are being implemented all over the world.

Many fisheries scientists, economists, and government agencies have bought into the concept of catch shares, including the US National Marine Fisheries Service (NMFS). Economists often think it's a good thing because under "rationalization," the profits of companies increase, and profit is how economists perceive value. From the scientists' perspective, they think fisheries could be managed more efficiently under the catch share system and that there may be benefits toward conservation.

In many fisheries, too many boats fishing for too few fish often leads to a "race for fish," sometimes called "the fishing Olympics." Government agencies place catch limits or effort limits on fisheries, generating intense competition among fishermen for the resource. They are forced to fish in unsafe conditions, sometimes causing loss of life, or to fish recklessly, often resulting in substantial bycatch of undesired species, which are then dumped at sea.

Much academic research (often receiving financial support from EDF, NMFS, or the foundations directly) demonstrates the favorable results of catch share programs. Well-designed catch share programs create safer conditions for fishermen and can help conserve fish populations. Individual fishermen know ahead of time how much fish they are allowed to harvest. They can spread their effort out over time, being more careful about their catch, and they don't have to fish in dangerous weather. Another selling point used to market catch share programs is the belief that if fishermen recognize that they "own" the resource, it's in their benefit to take good care of it, resulting in better management.

These advantages sound great, but are there downsides we aren't hearing about? Small-scale fishermen complain they don't have the political organization and funding support to mount competing media campaigns so their voices can be heard. When money influences public opinion so only one side is heard, there's social inequity.

According to catch share opponents, quotas to fish are often granted to industrial fisheries as a result of political influence and state economic policies. Since the quota share is based on past catch history, the bigger boats that caught the most fish in the past get the biggest quotas. Some conservationists point out that the fishermen responsible for overfishing the resource in the first place, thereby creating the need for catch shares, now get rewarded. In anticipation of catch share legislation, some fishermen caught as many fish as they could so their share would be maximized once the laws were enacted. One United Nations expert dubbed the practice "ocean grabbing."[13] Some scientists and watchdog groups argue that IFQ programs are not always effective in conservation or in reducing the amount of fishing power deployed to harvest the fish—they just transfer and concentrate it.[14]

Often the small quotas granted to artisanal fishermen are too small to support them, so the quotas are bought by larger companies, consolidating their grip on the fisheries. A single, vertically stratified corporation is more efficient for corporate profits than numerous small companies. Local fishing communities pay the price. Catch shares may be controlled by distant fleets, disrupting local economies. Increasing efficiency reduces costs by decreasing the number of fishing vessels, but a consequence of fewer fishing vessels is fewer jobs and less need for infrastructure support in the community. The fish becomes a commodity for corporate profit, rather than a resource that supports the local community.

Clearly, catch share programs have positive and negative effects. The negatives appear to mainly affect small-scale fishermen. How can we find balance?

Developing Solutions

Although many problems confront traditional small-scale fisheries, there are success stories as well. Fishermen and communities are banding together to save their fisheries. A trending locavore movement known as Slow Food, based on consuming locally grown/harvested food, helps support local fisheries.[15] As well, there have been innovations in cultivating fish, and progress has been made in restoring damaged fish populations.

In 1984, a group called the International Collective in Support of Fishworkers formed to counter the increasing emphasis of the FAO on commercial and industrial fisheries. In recent years FAO has shifted emphasis back to protecting small-scale traditional fisheries. Fishing cooperatives and community-based harvest programs have been created that hold promise for preserving traditional fishing practices.

Similar to the organic movement in agriculture that is now several decades old, a growing number of people want to know that the fish they consume were harvested responsibly and are healthy to eat. There are numerous certification and traceability programs for fish in the market to help shoppers decide which product to buy. There is also a growing movement to support local food sources. Cooperatives on the East and West Coasts of the United States specialize in brokering local catches "from sea to table," sometimes with fishermen selling their harvest directly to consumers. Such programs stress the intimacy of eating wild fish harvested from the ocean.

Aquaculture is envisioned as a solution to the problem of feeding a growing world population and providing an important source of protein in the future. As of 2012, about one billion people in the world suffer from hunger. Although the production of marine capture fisheries

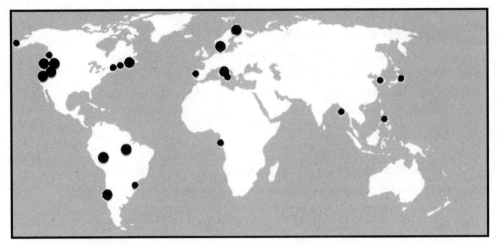

Figure 2. Map showing locations of the stories in this book, including the main chapter stories (large circles) and the shorter ones at the ends of the chapters (small circles).

has leveled off, in the past thirty years the production of aquaculture has increased at a rate of 8.8 percent per year, far outpacing human population growth.

Farm-raised fish have taken up the slack left by a fully exploited marine resource to provide protein to a hungry world. Cultivated fish were worth $119 billion in 2010. Over six hundred aquatic species are raised in captivity. Thus it appears the "era of expansion" of capture fisheries has been supplanted by the "blue revolution" of farmed fisheries. But are these farmed products healthy for people and the environment?

This book is a collection of stories about fishermen—their histories, the places they work, the problems they face, and the fish they catch (or raise). Most of the fishermen in this book fall under my classification of new artisanal. Several stories are about indigenous fishermen. Their history is important here because the story of indigenous fisheries precedes that of today's traditional fisheries, and may predict their futures. The narratives are based on interviews with fishermen, community activists, conservationists, scientists, and seafood growers. I have included several stories about small-scale aquaculture in this book because as a fishery enterprise, fish farming is now too important to ignore. Most of the stories are about the sea, but a couple of case studies are taken from inland waters. The setting doesn't matter as much as the theme. These chronicles zoom in to an individual level, giving a fine-grained view of the larger issues in the world's fisheries—too many fishermen with too few fish, conflicts with other resource users, loss of fishing rights, and degraded habitats. These are mighty challenges, but my outlook is optimistic—with good science, a multidisciplinary/cultural approach, and the perspective history gives us, we can make wise decisions for robust fishing communities and healthy oceans. Many of the fisheries presented here do just that.

FOULED FISH

What I am mourning is perhaps not worth saving,
but I regret its loss nevertheless.
John Steinbeck

Too few fish for too many fishermen. Too many fishing boats. Vastly modernized fishing technology. Degradation of habitat by development and pollution. Competing alternative uses of the water as a resource rather than a living space for fishes. For small-scale fishermen, these situations represent a one-two punch that is hard to withstand: declines in coastal fisheries production and overfishing. There is a new menace in the path of survival: multinational entities who want to privatize public natural resources for their exclusive use. If the wild fish disappear, it will not affect some of these corporations, because they can replace the wild fish in the markets they own with farmed fish that they grow on giant ocean ranches. For them, the bottom line is not subsistence or a decent living wage, but profit. And the impact is huge.

In chapter 1, the artisanal fishery in Italy is struggling against overfishing, competition from industrial trawlers, and degraded habitat. The population of fishermen is aging and disappearing. Chapter 2 describes the collapsed population of the Vosso River salmon, and the complete loss of the fishery in the face of environmental degradation and use of the wild salmon habitat and public waterways by commercial fish farms. Chapter 3 portrays the collapse of offshore fisheries in Chile that has coincided with privatization and the granting of exclusive fishing rights to four companies owned by several powerful families.

1: The *Giovanni Padre*

THE SUN SETS ON SMALL-SCALE FISHERIES IN THE GULF OF NAPLES

It was early morning when an autumn breeze rose off the Mediterranean Sea and pushed the curtains into the stillness of my room. Outside, lights blinked on the horizon from small boats that were returning from pulling their nets. A storm rumbled in the distance as layers of thunderheads clashed in the sky. Backlit by lightning flashes, the clouds looked like boxing Javanese shadow puppets. I watched as the boats drew close and deftly weaved their way into the channel. Later, down at the dock, fishermen handpicked the catch from their nets into white buckets. Then they dumped the contents onto a marble countertop in the public market area. A few chefs from local restaurants and several housewives gathered around the display of fish and exchanged friendly banter. The fish were sold individually for the larger species like the rare small tuna, or by the handful in small boxes for the tiny anchovies.

This is the rhythm of Lacco Ameno's harbor on the island of Ischia. The ebb and flow of the fishermen, the passing of storms, changing seasons, advancing years. But this ancient way of life is disappearing. Most of the fishermen of Ischia are gray-haired now, in their sixties or even older. A generation of young fishermen—apprenticing to learn the way of the sea and how to catch the fish—finally to replace the old men is lacking. They are rare at the dockside; the lure of safer and more lucrative professions has pulled them away. They've become tourist guides and taxi drivers on the island, or left for the promise of riches to be found in Naples, Rome, or maybe America.

In 2011, I traveled to Ischia and learned about problems that traditional small-scale fishermen face: depleted fish stocks, mostly from

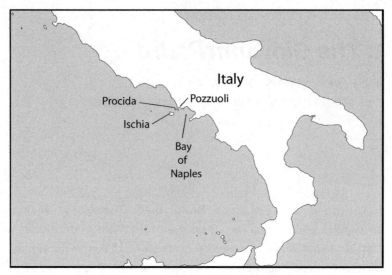

Figure 3. Location of the island of Ischia in the Bay of Naples, Italy.

overfishing; increasing competition from larger, more modern vessels; constricted areas where they can set their nets; an aging population of practitioners; government programs to grant fishing rights to a privileged few; and the struggle against the political power of big industry. These changes are happening not only in the Mediterranean, but all over the planet: Norway, Chile, Africa, the United States of America—you name it.

Ischia lies in the Gulf of Naples and is the largest of the Phlegraean Islands. Greek legend tells us that Zeus had mythic battles with his monstrous enemy Tifeo (Typhon) near here. Tifeo had some mean seeds sprouting from his loins. He fathered a pride of nasty offspring, including Cerberus, the three-headed hound guarding the passageway to the underworld; the eagle of Prometheus, the Chimera, the Sphinx, the dragon of Colchis, and the Hydra of Lerna, among many others. When Zeus finally killed Tifeo, he buried him in chains at the foot of Pithecusae, now known as Ischia. The rumblings from Tifeo's grave are blamed for the many earthquakes that shake Ischia's residents.

The island was colonized in the eighth century BC by the Greeks. It has swapped hands like the queen of spades in a game of hearts, having been conquered over the centuries by the Roman Empire, Spain, France, and England. The Ottoman Turks and Barbary pirates—including the infamous Redbeard—raided and pillaged Ischia. Finally, the island was annexed by the government of Naples.

Ischia isn't known for its tradition of fishing. Rather, it is green and agricultural. The island's rich soil grows grapes and tomatoes exuberantly. Cuisine-wise, Ischia is recognized for its wine and rabbits. The sea was used to ship the island's wine to the mainland, and to a lesser extent for subsistence fishing and to feed the local market. More recently, the fishermen of Ischia received a boost as the island became a tourist destination and the visitors wanted seafood on the menus of upscale restaurants.

Neptune's Garden

One of the old men off-loading his catch in Lacco Ameno's harbor had a weather-beaten face and kind eyes. He wore a faded black cap and a baggy blue sweater, and his trousers were tattered. His name is Paolo Vespoli. He fished alone on his small boat, the *Giovanni Padre*. (On this island of Ischia, there is a tradition to name your boat after your father.) Paolo dumped and quickly sorted his catch on the community table for waiting buyers: a meager bucket of small silver fishes for a night's labor.

Morning at the harbor smelled of fish and diesel. I tried making small talk with Paolo using a few phrases I'd translated and memorized.

"Io sono un amico di Lorenzo di America," I said. I am a friend of Lorenzo in America.

"Ah, Lorenzo," Paolo responded with a smile.

"La cattura del pesce è piccolo," I said. The catch is small.

"Sì, piccolo."

My colleague at home, Lorenzo Ciannelli, had told me to look for Paolo. Lorenzo's father had been a fisherman on Ischia. I was trying to talk my way onto Paolo's boat for a fishing trip.

"Is that your boat? Err . . . vostra barca?" I asked, realizing that I was running out of Italian words to make small talk.

Figure 4. Fishermen unloading their catch in front of prospective customers, Lacco Ameno, Ischia, Italy. Photographer: K. Bailey.

"Sì, il *Giovanni Padre*," he answered.

My broken Italian began to crumble. Spanish and English words got tossed into the mix.

"I want to go . . . puedo . . . andare pesca," I said, panicking.

He looked perplexed and said, "Non capisco" with a shrug, arms raised and palms opened toward me. Gulls chortled and cawed in the harbor.

There was a suggestion of finality in his gestures, making it clear that the conversation was over. I think he knew what I wanted, but maybe I would disrupt his solitude at sea. I would be a bother—after all, I was just a tourist, one of many here, and he was trying to make a living. He turned from me to attend to his business.

Later I learned it was illegal for Paolo to take me on his boat. To do so, we'd have to get permission from the Italian bureaucracy, and by the time it arrived I'd be long gone.

The next time I saw Paolo I had a friend along to interpret, and now he talked freely with me. I asked about his history fishing. Paolo told me that he has always had a passion to fish, even when he was six or seven years old and would fish with his father. One day the seas were rough, and his father left home without him. Paolo remembered how upset he was about being excluded; he never let that happen again.

Paolo said that fishing is different now. There are fewer fish than when he was young. According to him, the ecosystem has changed. Some species that he used to catch are no longer there. Big trawlers and seiners have improved their technology. They have increased the size of the nets they put in the water and are catching more and more fish. In his small boat, Paolo now fished in canyons and rocky reaches where the big trawlers can't go.

According to Paolo, the development of the harbor's shoreline is another reason the fish aren't abundant anymore. Concrete bulkheads armor the shoreline in many areas. Pollution and development have killed the seagrasses that provide a nursery to juvenile fishes. The best harbors for boats also tend to be prime habitat for young fishes.

It wouldn't appear that overfishing should be a problem here; there are few big fishing trawlers harbored in the bays of Ischia. The boats are mostly wooden and small, carrying one or two men. The fishermen lay out a floating wall of fine netting—a type of gillnet called a *tramaglio*—and the fish swim into them and get caught in the mesh. Then the fishermen haul the nets back by hand or with a small power reel, and laboriously remove each individual fish from the netting. It is tedious work. Some of the boats set and retrieve their nets twice a day.

Ischia is surrounded by a marine nature reserve called Regno di Nettuno, or Neptune's Kingdom. The area of the reserve is only about one hundred square kilometers, or about the size of Nantucket. The activities of fishermen are controlled in the reserve. Since their small boats have a limited range, the men used to fish close to the island.

Figure 5. Small fishing boats in the harbor of Lacco Ameno on the island of Ischia, Italy. In the background are large yachts competing for space in the harbor. Photographer: K. Bailey.

But now that the fish near the island are depleted, some of the small boats are forced to venture longer distances to find fish. The idea of the reserve is to replenish the fish, but it makes life difficult for the local fishermen.

The neighboring island of Procida is better known for its fishing tradition. This island is small and barren. The sea has provided for the people living there. Since Procida is just a few minutes from Naples by boat, a ready market of seafood-loving Italians waits nearby. Fishermen of Procida have invested in bigger and better ships in order to catch more fish and satisfy the market's demand. Instead of fishing with the small, passive tramaglio used on Ischia, the fishermen from Procida use their more powerful ships to drag bigger nets through the water, trawling the fish from the bottom of the sea. But the bigger boats are more expensive, they cost more to operate, and their owners have to take out larger loans. They have to catch more fish to pay their bills. It is a vicious cycle.

Fishermen like Paolo on Ischia say that the big trawlers from Procida have overfished the seas surrounding the islands. They also say that sometimes the fishermen from both islands may engage in illegal fishing to make ends meet. The Procida boats are faster, their range greater. At night they may sneak into the marine reserve where fish are plentiful, or by day they set their nets on the edges of it. The fish don't recognize the park boundaries and stray into the danger zone, where they get scooped up. There is conflict with the small-scale traditional fishermen when the fish are in low supply because the trawlers can outcompete the small boats for the harvest. Then, just as in other parts of the world, politicians enter the fray and enact policy based on the muscle of money and influence. This result is really no different in Italy than it is in Alaska or Chile, just on a smaller scale.

The struggle for survival of the small traditional fishermen of Ischia is symbolic of the global battle between the large industrial fishing companies and traditional fishermen.

Industrial fishing grew rapidly after World War II. Global fish catches more than doubled over the following two decades, fueled by government subsidies to reconstruct ships that had been destroyed by enemy bombs and torpedoes. After the fleets were rebuilt, subsidies continued (now for fuel and marketing), giving the big companies a competitive advantage in the global marketplace. The rebuilt Russian and Japanese fleets roamed the world's oceans in search of more and more supply. They were joined by the Spanish, Americans, Norwegians, Cubans, Chinese, Koreans, Germans, Poles, and others.

After thousands of years of small-scale traditional fishing, the wave of industrial change began to wash over and suffocate the small-scale fishermen. Then, in 1953, the Birds Eye division of General Foods announced the production of frozen fish sticks. A little later came frozen fish fillets. At this point the fast food industry synchronized with the fishing industry, and they all moved into high gear. The harvest of fish from the sea accelerated like it was an industry on steroids.

Dinner at Dada's

At the harbor of Lacco Ameno, a man shopping the day's catch was Vincenzo Caputo, the owner of a private restaurant in the little village of Casamicciola, just up the road. He chatted with the fishermen as he looked over the catch, and picked out some calamari and lampuga (*Coryphaena hippurus*; the same species known as mahi-mahi or dolphinfish in the United States). He bought some anchovies from Paolo. Vincenzo is about fifty years old, svelte, very tanned, with thick blond hair. His smile is large and easy, and his white teeth stand out. This day he wore a crisp white shirt unbuttoned far enough down to reveal a gold chain and the lack of a tan line.

Vincenzo's restaurant, called Dada's, serves food only by appointment to a collection of friends (and their friends) on the island. My friend Maria, a marine biologist living on Ischia, arranged for my dinner there. The restaurant occupies the bottom floor of a former warehouse, with several rooms. Entering the restaurant, I walked by an open kitchen where Vincenzo presided over dinner like a conductor at the podium. From there he sees and greets everyone in his audience. Vincenzo danced across the floor of his kitchen in constant movement. He waved utensils and bunches of basil in the air as he prepped for dinner, chattering enthusiastically in Italian about what he was doing. There were platters of butterflied anchovies, calamari, basil, and tomatoes. He prepared them as if he were fine-tuning an orchestra.

The anchovies were headed, gutted, and stripped of their bones with skill. Vincenzo marinated them in fresh-squeezed lemon juice for twenty minutes. Then he drained them, drenched with extra-virgin olive oil, and mixed the fillets with diced garlic, peperoncino (also known as *diavolicchio*—a word that's best released from the tongue along with a flamboyant gesture of the hands), and parsley. They were served as the antipasto, the hors d'oeuvre.

Vincenzo skinned the calamari under water and cut the flesh into rings. He heated the shellfish for a minute to sweat them of water. When the meat was white, he poured off the excess liquid. Then he added the squid to hot olive oil with peperoncino, garlic, basil, and parsley, and fried it. He salted to taste and then added some white wine. The mix was further cooked for ten minutes. At the rear of the stove the pasta was boiling, setting a gurgling rhythm. A plume of steam rose from it. Freshly crushed tomatoes were added to the saucepan, then simmered for another ten minutes with some of the starchy water from the pasta. Finally the calamari was added to the drained pasta with black olives and capers, twirled on a fork against a large spoon to make a ball, and set on the plate as the primi piatti. Some extra calamari mix was spooned on top. The squid was soft and succulent. Delizioso.

The plan was to fry the lampuga with laurel leaves, but at the last minute Vincenzo changed his mind. The meal would be too heavy—more like a requiem than a cantata. Instead, he filleted and skinned the fish, cut the fillets into very thin strips, and marinated them in lemon juice for five minutes. Then, after squeezing the lemon juice out, he drizzled the lampuga with extra-virgin olive oil and served it over a bed of arugula (rucola).

The wood-fired stove was stoked and ready for the pizza. Vincenzo constructed a simple Margherita pizza of fresh basil and crushed tomatoes, topped with a mixture of mozzarella di bufala (water buffalo), some more cheese that he had smoked himself, and of course peperoncino.

The courses were linked together by a free-flowing glissando of local white table wine, and capped with homemade limoncello. Having overeaten, we finished off the evening with a stroll on the cobblestone road along the waterfront—seemingly accompanied by all of Casamicciola—and finally with an espresso at a beachfront café.

Some of the guests that night were members of a local marine conservation organization, and there were several marine biologists, including Maria. They talked of the artisanal fisheries and noted with irony that the "slow food movement," or the practice of using locally caught or grown food, is a traditional lifestyle affordable only to the relatively affluent. Artisanal producers and fishermen get a good price for their organic and local food, but it is grown and caught in limited amounts. The artisanal fishers can hardly afford to eat what they catch and sell. It's cheaper to buy frozen fish sticks made from pollock that are caught in the Bering Sea and processed in China. So oftentimes they sell their fish and buy industrial fast food for less, keeping some money in their pocket. Given the global marketplace and increasing human population, fresh fish are getting harder to find and more expensive, even for the relatively affluent.

With the rise of industrial fishing, restaurants started buying frozen fish; even in Italy, many of the larger tourist restaurants now use frozen fish. For one thing, it is a lot of work to go to the market every morning and buy the day's catch. Then it has to be cleaned and scaled, or cut and filleted. Many customers expect the fillet to be deboned. When you serve many customers—say, more than fifty mouths—to do all those tasks is just too much work. Besides, the small local artisanal fishermen don't catch enough of any one species, nor do they provide a dependable and continuous supply. Diners on holiday don't like to be told that their selection from the menu is not available: "How about a nice fillet of farm-raised Chilean Atlantic salmon instead?" Never mind that Chile's coast isn't on the Atlantic Ocean and the salmon are raised on three kilograms of industrial-caught "forage" fish to make one kilogram of salmon. The anchovies and sardines that Paolo caught and Vincenzo served are fed in greater quantity to the salmon. It's the global marketplace at work.

The World Bank Pushes for Efficiency

Seeing what is happening to the traditional fisheries brings forth a mixture of sadness and introspection. I ask myself if my grandchildren will be able to experience the enchantment of watching the fishermen take their boats to sea as I did? Granted, in high-seas fisheries, big boats are much safer in the treacherous waters. For sure they are more efficient and keep prices down for consumers. They help provide protein to hungry people on land. But the loss of the independent fisherman, and a way of life that intimately ties us to the sea, is mournful. Workers on factory trawlers often spend sixteen hours a day, seven days a week belowdecks. They stay out fishing until the hold is full. There is no rhythm with the sea. This isn't the idyllic life at sea they'd imagined.

About thirty thousand of the thirty-five thousand boats fishing in the Mediterranean Sea are still small-scale traditional fishers. Over the two decades from 1990 to 2010, the spawning biomass of commercial fish stocks in the Mediterranean declined by an average of about 75 percent.[1] In response, the European Parliament decided that action was needed. They proposed in the Common Fisheries Policy to implement an individual transferable quota (ITQ, or catch shares) system in this region. Bureaucrats envision ITQs as a panacea to overfishing. The World Bank was firmly behind the scheme.

Since the right to fish can be traded, leased, or sold, there is a political force at work. The large companies have the financial strength to purchase quota, and this has the effect of cannibalizing the small participants, consolidating wealth, and building political muscle. Aging fishermen like Paolo will eventually cash out and sell their quota to companies. Especially if the fish stock they harvest undergoes a decline. In the old days they would switch to another species, but now they have to buy some quota of that species as well. What's missing from the equations of economics underlying the system is the unpredictability of human behavior.

With consolidation of the fisheries, fewer boats are needed to catch the total quota. Now, fewer fishermen can do the job working longer hours. While consolidation increases efficiency, thereby yielding greater profitability for the survivors, the consequence is fewer jobs in fisheries and in the surrounding community that services them. The increased profits of fishing entities—by now often corporate—enable more political investment. Fishermen are replaced with lobbyists. Increasing political investment leads to more power in a corrupt cycle.

The ITQ implementation looks like another nail in the coffin of the fishing tradition. Although economic profitability and political influence will be greater for some, the surrounding communities will be poorer. Wealth gets concentrated. There is a trade-off between supporting healthy fishing communities and extracting more profit for a few owners and financiers.

What can we do? There is no global panacea to the overfishing problem, but only local solutions that adapt to unique and complex situations. We can enact quotas that are nontransferable, save some for small-scale fishers, and restrict coastal areas for the use of local and traditional fishermen; in some regions, "community-based" quota systems are being tried. We can preserve markets for local fisheries by promoting the accurate labeling of fish and locally caught products. In Italy, there is a unique alliance of the "slow food movement" and traditional fisheries, like the linkage between Paolo and Vincenzo. This movement is growing fast and globalizing. Sometimes they are called "dock to dish" fisheries, or community-supported fisheries. Maybe the slow food movement offers hope for salvation of some artisanal fisheries.

On our stroll along the waterfront after dinner at Dada's, another storm loomed on the horizon. The atmosphere was heavy with a sense of foreboding. The marine biologist Maria looked toward the gathering clouds and remarked about how the biodiversity of the dinner plate in Italy is shrinking. With a hint of sadness, she noted that the foods—and especially the variety of seafood that she ate at her grandmother's table as a young girl—aren't available anymore. Traditional ways of sustaining ourselves are being swallowed up by industrial efficiency, profitability, and affordability. The population of fishermen is aging as catches diminish, and their political force is waning. Overhead, the storm grew closer and the clash of titans played out across the sky. The thunder suggested the anguish of the old gods. A way of life dating back to their reign of the heavens, an ancient craft, was passing.

Postscript

Four years later, I returned to Lacco Ameno. Now there were only three fishing boats in the main harbor where before there were six. The *Giovanni Padre* wasn't among them. Luxury yachts, crewed by men in white uniforms, replaced the working vessels. Newly installed plastic tables and barbecue stations decorated the walkway on the jetty.

I noticed a strange daily migration of the yachts. In the morning they cast off and moved several hundred meters up the shoreline,

Figure 6. Fisherman Paolo Vespoli in the harbor of Lacco Ameno on the island of Ischia, Italy. Photographer: K. Bailey.

where twenty or thirty of them anchored off the beach of a resort. The crew rowed the yacht people ashore in skiffs. They rented spaces on the private beach to sunbathe and swim. These beaches are actually public, but the city has leased them to a private company in Naples that re-rents space on the beach back to the people—almost like a feudal tenant system. I watched one of the skiffs return to the beach to pick up the shore party. It was too big to row, and since motors aren't allowed in the swimming area, the crew jumped in and pushed the boat ashore. In the late afternoon the yachts motored back to the harbor. The scene could have been taken from a Fellini movie.

Through my friend Lorenzo, I arranged to meet with Paolo once again. He had sold the *Giovanni Padre* and said it was moored nearby, in the harbor of Forio. Paolo was tanned and sported a bright white shirt, unbuttoned enough to display a gold chain with a Saint Christopher's medal. He said he was getting too old to fish, but also that the arbitrary and bureaucratic regulations about fishing were getting to him. He even got ticketed for taking his son out fishing with him. Then he told about how he got cited for selling a large catch of mack-

Figure 7. Drawing of the modern version of the fishing technique called *pesca d'ombra*. The technique uses gear called *cannizzi*—plastic palm fronds attached to floats to keep them from sinking and anchored with lines tied to concrete blocks. The fish are attracted to the fronds and then captured with nets. Artist: Mattias Bailey.

erel in the market—something he'd done all his life, but on this occasion it was a violation of permitting rules. He was put in jail for a few hours, and he'd had to hire a lawyer. He gave up fishing. Now he spends his days around the harbor and mends nets.

Pesca d'ombra

Paintings on some ancient Roman urns and tile mosaics depict fishermen catching the lampuga using a fishing technique called *pesca d'ombra* ("fishing of shadows"). In this way of harvesting, the fishermen use a type of gear called *cannizzi*, which are palm fronds attached to floats to keep them from sinking. They are anchored with lines tied to stones. The shadows attract the fish. When enough fish have gathered around the cannizzi, the fishermen in boats surround the fish with a net like a seine and haul them in. Similar methods of fishing are still being used in Sicily and other places in the Mediterranean, but now the palm fronds are artificial, made of plastic leaves or frayed polypropylene rope, and the anchor is made of concrete.

2: The King Is Dead

THE COLLAPSE AND RESURRECTION OF VOSSO SALMON

High above Norway's Bolstad Fjord, perched in a cliff-top hut, Helge Furnes watched with the intensity of an osprey for a monstrous salmon to enter his trap of netting laid out on the river bottom. A system of ropes and pulleys connected the net to a pair of heavy stones that he had hoisted up to the bottom of the hut. When he saw fish enter the trap, Helge yanked the end of the rope, setting off a chain of reactions accompanied by a cacophony of sounds that was music to his ears. The stones dropped toward the ground; their falling weight whipped up lines that were tied to the corners of the net; pulleys whirred; and the trap closed, capturing the fish within.

This method of fishing for salmon on the Bolstad Fjord, known as *sitjenot* ("sitting net"), dates back at least 150 years. Helge's father fished for salmon this way. But the fish are mostly gone now, and Helge no longer fishes the sitjenot. The huts sit, still perched on the cliffs or on poles above the fjord, like ghostly sentinels waiting for the return of the salmon.

The Vosso salmon is the stuff of legends. Once these were the largest Atlantic salmon in the world, based on average weight. Some behemoths tipped the scales at more than thirty-six kilograms. Catches of salmon in the Vosso River system were relatively stable for hundreds, maybe thousands of years, averaging about twelve tonnes per year. Then suddenly, in the late 1980s, the numbers began to nosedive. The fishery collapsed in 1991 and was closed in 1992. For all practical purposes, the wild stock of Vosso River salmon went extinct. The "king of fish,"[1] as they had been called, were no more.

The decline of Atlantic salmon was not unique to Vosso—similar

Figure 8. Sitjenot fishing hut in Bolstad Fjord, Norway. When observers in the hut see salmon entering the trap in the water below, weighted stones attached to ropes and pulleys are dropped, which haul up the ends of the net, capturing the fish. Photographer: K. Bailey.

decreases in other European salmon stocks occurred at about the same time. But what happened in Norway after the collapse was unique, and probably couldn't happen anywhere else. That's because of the cultural significance of the Vosso salmon and the financial resources available to the Norwegians. These factors were compounded by an ironic turn of events.

When oil was discovered in the North Sea in the 1960s, Norway struck it rich; enormous wealth was generated by sovereign control of the oil fields, leading to a system of taxation on oil extraction. Furthermore, there was direct production and sales by the state-owned oil company. With the inflowing revenue, the Norwegians invested generously in infrastructure, social welfare, and research, including the development of a fish-farming industry.

When the wild salmon populations suddenly declined, Norway's government funded conservation efforts. In the late 1980s, fueled with oil revenues and extra financing from hydropower companies, scientists created a national living gene bank in Eidfjord, a village that sits on the edge of the Hardangerfjord, to preserve the genetic diversity of salmon. Here, live fish from different wild stocks were kept in land-based fish farms. It was a salmon zoo. Among the preserved stocks was the Vosso salmon. The foresight to preserve its gene pool of natural fish populations is unique to Norway.

In 2000, fisheries scientists teamed up with the national government, local governments, universities, and the power and salmon farming industries to revive the wild Vosso River salmon. By then, about 70 percent of the salmon entering the river to spawn were escapees from the salmon farms, and the remaining "wild" fish were likely hybrids. The scientists believed that this was a "now or never" situation: the remaining fish were unlikely to survive because their specialized local adaptations had been diluted by breeding with farmed fish. A program was initiated to plant the river with smolts of the native Vosso stock cultivated from the Eidfjord farm.

Now, fifteen years after the rehabilitation program began, native Vosso salmon are back in the river, migrating to the ocean and returning to spawn as they once did. The problem is that the naturally spawned offspring aren't surviving. The survivors that return from the ocean are those that were grown in hatcheries, which were hauled as young fish to the ocean in big cages and released there. Something is still wrong in the Vosso River system, and the mystery has generated a lot of controversy.

A Trip on the Fjord

The Vosso River is about eighty kilometers east of Bergen, on the west coast of Norway. The strong, deep flow of the river carves through rocky cliffs and winds its way past green pastures dotted with red

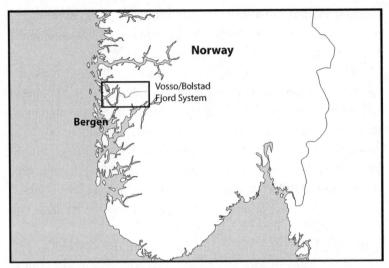

Figure 9. Location of the Vosso River and Bolstad Fjord system, Norway.

barns and yellow houses. It is a picturesque setting. Voss is a sleepy village, maybe best known as the birthplace of legendary American football player Knute Rockne. A little farther downstream, where the river escapes from the mountains and widens, it empties into the Bolstad Fjord, which then joins a complex inland fjord system spreading one hundred kilometers to the ocean.

The Voss region was settled about three thousand years ago by hunter-gatherers. The evidence for salmon use by the prehistoric men is about as old as the first recorded occupation. Later, farming took hold in the fertile valley and Vikings inhabited the area; there is a Viking burial site near the town of Voss. In AD 1023, King Olaf V converted the region's pagans to Christianity, just around the end of the Viking era.

My hosts at Vosso were Drs. Knut Vollset and Bjorn Barlaup of Uni Research,[2] a consulting offshoot of the University of Bergen. I knew Knut from nearly a decade before, when he spent six months in my lab as a visiting graduate student. Bjorn, Knut, and Helge (the ex–sitjenot fisherman) took me on a boat tour of the river system to learn about the Vosso salmon recovery effort.

The waters of the fjord on this autumn day were calm and smelled faintly trouty. We stopped to climb a long, rickety gangplank up to one of the sitjenot huts that sat high above the river, overlooking a deep pool below. I tread cautiously as the old timbers groaned in the wind with the sound of wood rubbing against wood. The huts aren't used for commercial fishing anymore; scientists now employ them to study the salmon. (They told me about one of their young unmarried scientists who had spent several lonely years in a hut counting salmon as they passed through the maze of nets below. His disheartening adventures in finding love while residing in the shack were a topic of mirth in the local community. The arrivals of young women at the scientist's hut were telegraphed across the riverbank and followed almost as closely as the passing of salmon in the river.)

After our tour of the river, we visited Helge's farm. Helge was about to turn eighty years old. His body was lean, and his eyes sparkled with enthusiasm. He lived with his wife, Kirsti, on the roadless side of the fjord in a neatly kept yellow house perched on the steep hillside. They raised sheep, a skittish breed with a crazed look in their eyes. There was an old cable-and-tram system to cross the fjord in an emergency when the waters were too rough or ice too thin to make the trip by boat or on foot. Stepping onto Helge's farm was a visit to another era, a privileged trip back in time. After we scrambled up a steep hillside to his house, we took off our boots and coats in the mudroom and entered a warm cozy kitchen. We sat at a four-person table next to a window overlooking the fjord, while Kirsti served us fresh blueberries and waffles for an afternoon snack, washed down by cups of black coffee. Unable to keep up with Bjorn and Knut, I limited myself to four servings. Kirsti bustled about the kitchen; she seemed disappointed by my performance and encouraged me to eat more. In the meanwhile, Helge talked.

The words flew off Helge's tongue as if they'd been waiting for just this opportunity to escape. His eyes widened with excitement as he chattered in Norwegian. Knut acted as my interpreter.

"I think there's a point to this, but he keeps diverting from the story to describe where he finds different berries," Knut said. As Helge rambled, he touched on the berry patches as if they were cairns marking the path of his narrative.

"Now he's telling a story about his uncle Sivert," Knut said. "It was

Figure 10. Helge Furnes in his boat on Bolstad Fjord, Norway. Photographer: K. Bailey.

the middle of winter, maybe about 1890. February. Sivert was sickly when he was born and his parents thought he was going to die, so when the baby was just a few days old, they swaddled him and began to hike the icy trail over the mountains to the nearest church in Stamnes. He needed to be baptized before he passed on. On a steep part of the climb, Helge's grandmother slipped on the ice. The baby fell from her grasp and slid down the hill on his backside like a sled. Whoosh. Barely missing trees and rocks. The bundle came to a stop just short of the edge of the fjord. They retrieved the bundle and continued on to the church."

Sivert didn't die that winter after all. Instead, he grew into a healthy man. But times were tough; the farm and fishery couldn't support another potential family, so Sivert emigrated to the United States when he was seventeen years old. It's said that there were so many immigrants in America from the region around the Furnes farm, at one time they made up a whole football team in Iowa. Sivert returned to the farm as an old man in the 1960s and was often seen sitting by the fjord near the salmon nets, singing old folk songs from his youth.

In a moment of reflection, Helge gazed beyond the steam rising from his coffee. He said that people would often remark how hard his life must be on the edge of the fjord, especially without a road. He pointed out a resort-like lodge up on the hill behind him that a millionaire from Bergen had built. On the weekends, his neighbor would helicopter in to this country manor.

"My life can't be so bad. After all, people like my neighbor work like dogs for weeks in the city in order that they can enjoy a few days living here," Helge said. His eyes twinkled.

Returning to the topic of salmon, Helge pointed out that in the old days, the huts were carefully placed to put the nets in the path of the salmon. The fishermen stained the cliffs white with lime to trick the fish into thinking there was a waterfall above, intending for the fish to veer toward it and into their traps.

In the old days, sitjenot fishermen were proud of their skill. There was a friendly rivalry among them; each would keep an eye on the neighboring huts to see what they were catching. When fishing was good, a big catch was up to fifty fish. The local fishermen also expertly guided tourists from England to fish for salmon in the Vosso River, as the giant fish were famous worldwide and prized among tweedy fishing connoisseurs.

The Decline of Salmon

The sudden demise of the Vosso salmon stock after thousands of years of existence points a finger directly at the human activity disrupting the relationship of salmon with its natural world—although who specifically is responsible is a subject that generates controversy. The situation in the Vosso River ecosystem is complex; and as in other complex systems, the human instinct to engineer and tame nature in the Vosso River for our own purposes has had unforeseen consequences on salmon and other species.

A steady decline of Atlantic salmon across the Atlantic Ocean tracks the progress of industrial development in salmon habitats. Wild stocks have disappeared from over three hundred river systems in Europe and North America. Salmon are endangered in another

Figure 11. The father of Helge Furnes with Vosso River salmon. Used with permission from Håkon Furnes.

one-third of their native rivers. But in the three decades from 1970 to 2000, the population decline has accelerated, with the total catches falling by 80 percent across the salmon's range.[3]

Statistical results from models covering some sixty river stocks over about thirty years show that salmon populations are influenced by changes in the freshwater environment where they spend the first two years of life, and by variations in the ocean where they live for another two to four years. In particular, researchers have found that on a broad scale, salmon survival depends on interactions of ocean temperature with the presence of fish farms, and also with the presence of hydroelectric development in freshwater.[4] But on a finer scale, the conditions influencing the health of salmon populations vary from river to river.

Salmon are doing somewhat better in Norway than in most other countries, but even so, their declines are noteworthy. There are 667 rivers in Norway with salmon. A recent report, considering 180 populations that make up about 95 percent of catches in Norway, classifies only 36 percent of them as healthy. From 1983 to 2014 there was a 50 percent decline in salmon abundance across all Norway. By looking at geographic trends in salmon populations, Norwegian fisheries scientists see some interesting patterns emerge. But the patterns they observe vary considerably from place to place. The greatest declines, nearly 70 percent, have occurred on the west coast of Norway where the salmon farms were more developed.[5]

The arrays of salmon farm pens in the water trouble fisheries biologists, since they contain fish in very high densities—up to two hundred thousand fish per cage. At such high densities, the transmission of diseases and parasites can threaten wild stocks. Farmed fish can escape, breed with local wild fish, and impair the fitness of offspring to survive in local conditions. In the south and north of the country where there are fewer farms, the wild stocks have held steady.

Among the west coast fjord systems, scientists have noticed smaller-scale patterns with respect to fish farms and salmon health. The coastal stocks inhabiting areas with fewer salmon farms were doing better than those in the fjords where the farms were concentrated. Populations that originated deep in the fjord systems and which have to run a longer gauntlet of the farms, were doing poorly as well, yet another indication that the commercial salmon farms influence the well-being of wild salmon populations. Indeed, the Scientific Advisory Committee for Atlantic Salmon Management believes that the two greatest threats to wild salmon in Norway are sea lice from the salmon farms and the threat of escaped salmon from the farms mixing with the wild stock—it's been estimated that more than ten million farmed salmon in Norway have escaped into wild salmon habitat.

The annual production of farmed salmon in Norway is over 1.3 million tons, more than one thousand times greater than the harvest of wild salmon. It's a wealth-building industry that's been promoted by the Norwegian government, and the big farmers wield significant political power. The fish-farming industry has been keen to show that they didn't cause the decline of salmon in the Vosso. They point to other factors, such as predation on young salmon by sea trout, or supersaturation of the water with oxygen and nitrogen from the hydropower industry.

Indeed, though it's likely salmon farms play a large role, the diseases and parasites associated with them aren't the only factors causing problems for wild salmon. In the 1980s, the Vosso salmon was positioned at the confluence of a maelstrom of man-made factors that would influence the fish's fate, including acidification of the river system, hydroelectric development, increased siltation of the river bottom from road and railway construction on the riverbank, flood control, and loss of habitat. Changes in ocean temperatures may also affect survival during their two to four years of life at sea. All these things—and probably other as-yet-unrecognized factors—have had a cumulative effect on salmon survival over the past three decades.

The smolt-release program that began in 2000 was developed to revive the stock of native Vosso salmon, but the program doesn't involve just pouring fish into the river. Scientists have used the program to conduct experiments to learn about the Vosso ecosystem. Research on the fate of the smolts has begun to show results that are helpful to reveal the underlying causes of survival patterns. In particular, these studies have shown that smolts released in the river have a very low survival, which bodes poorly for a completely natural and self-sustaining run of salmon. Those smolts released at the coast have a much higher likelihood of returning to the system after their years at sea. Smolts released mid-fjord have an intermediate level of survival, while those released in the river itself have very poor survival rates. There seems to be a critical process at work within the fjord system.

We can speculate on the cause of the Vosso River salmon extinction, but it's largely a futile exercise to prove a single cause among the tangled web of factors. In the public debate, scientists hired to represent different industries and factions have different versions of what happened, and the data to demonstrate cause and effect are lacking. Some scientists believe that rather than point fingers, what we need to do is provide the healthiest environment possible for the stock to be productive, and to mitigate human-induced effects where scientific studies indicate there may be problems.

Already, progress has been made to temper the effects of acidification and variations of river flow on salmon survival. Liming of the rivers has returned the river to a normal level of acidity, but it doesn't seem to have had much effect on smolt survival. The hydroelectric industry has altered its methods of operation to improve conditions for smolt survival. Flows are regulated to maintain at least minimum requirements, and habitat lost in flood-control engineering has been restored.

Given all this effort and attention, some might wonder why the Vosso River salmon is important to preserve. "We expect African countries to preserve lions and elephants. Shouldn't a wealthy country like Norway preserve the Atlantic salmon?" Bjorn said. "It's a cultural icon, part of the national heritage, and is something important to the identity of the local people. If we hadn't done something, the world's largest Atlantic salmon, adapted to the largest fjord system of western Norway, would have disappeared without anyone knowing what happened—all within a decade."

What's more, if restoration efforts are successful, the Vosso River system could be a world treasure and a site for research and science outreach to demonstrate the complexities of how fish behave within the natural world and how humanity interacts with the rest of nature.

After coffee at Helge's house, we toured his boat shed. Inside there were several double-ended open-hull boats more than a hundred years old. Knut and Bjorn told me how active Helge had been in working to restore the Vosso salmon after the collapse, sharing his knowledge and helping out in the sitjenot traps the biologists now use to catch and measure the fish. They said that restoring salmon to the Vosso was his passion. It was apparent that the biologists treasured the fisherman. Helge's old nets hung from the ceiling as a reminder of times past.

We climbed into his skiff and he motored us across to the other side of the fjord for our drive back to Bergen. After Helge dropped us off, he zoomed across the fjord back to his home, leaving a broad wake behind him.

Later that evening as I looked over my notes, I thought of many other questions I wanted to ask Helge, but I was flying out the next morning. I thought it shouldn't be a problem—I could always return to Bergen for a follow-up interview.

A few weeks after I returned to Seattle, Knut wrote me a note. The air in the room became still as I read Knut's words. Helge Furnes had passed away. He was chasing his sheep in the steep meadows behind his house when his heart failed. An image of Helge scrambling after the sheep on the hillside remains in my mind. I felt blessed to have spent a few hours with him, catching a glimpse of his life with the great salmon.[6]

Farmed versus Wild Salmon

In 2014 about two and a half million tonnes of salmon, mostly the Atlantic salmon (*Salmo salar*), were grown in netted pens in marine waterways around the world. The number of salmon farms is increasing every year. A single Norwegian company grew about one-quarter of that total global production.

Salmon farming has been controversial in coastal ecosystems such as those of Chile, Norway, Scotland, and British Columbia, and caused problems for local wild stocks of salmon. A one-thousand-cubic-meter net pen holds up to ninety thousand fish, and there may be a dozen other pens nearby, resulting in unnaturally high densities. Waste from the farmed fish accumulates on the bottom, causing anoxia. Many people believe that the farms are directly responsible for declines in wild populations. The correspondence of fish-farm development and declines of local salmon populations in almost every place they've been deployed is hard to ignore. Often the pens are placed directly in the migratory routes of wild salmon stocks.

The farmed salmon in net pens are associated with dozens of diseases that can affect wild populations. In many locations, antibiotics and antifungal agents are applied excessively. The use of teflubenzuron to treat sea lice indiscriminately kills all crustaceans in the vicinity, including crabs, lobsters, and krill. Escapees from the salmon farms may breed with local wild fish, diluting the specialized local adaptations that have allowed the wild fish to thrive.

It takes about three to five pounds of low-value forage fish caught from the ocean to create a pound of salmon (on a wet weight–to–wet weight basis; other estimates in the literature at large are sometimes based on dry weight to wet weight). The dried pellets made from the forage fish to feed salmon have been linked with toxic antioxidants applied to preserve them during storage and transport. Some doctors and scientists advise against eating farm-raised salmon especially from countries where environmental standards are loosely regulated.

3: Ode to the Sea

CHILE'S TROUBLED FISHERIES

We are meager fishermen,
Men from the shore
Who are hungry and cold
And you're our foe.
Don't beat so hard,
Don't shout so loud,
Open your green coffers,
Place gifts of silver in our hands.
Give us this day
our daily fish.
Pablo Neruda, from "Ode to the Sea"

The Ley Longueira

The 787's landing gear thumped, jostling us awake. Seconds later, the plane's reverse thrusters blared an eye-popping roar to announce our arrival in Santiago, Chile, after a twelve-hour flight from Los Angeles. The clouds were lifting on what promised to be a fine autumn day.

The next morning we woke to the sound of songbirds chattering outside our window. I had a meeting scheduled with Juan Carlos Cárdenas, director of Eco Océanos, a nonprofit organization in Chile involved in fisheries conservation. I'd learned of his aim to repeal the Chilean fisheries law known as the "Ley Longueira." Cárdenas would be the first person of several I interviewed over the next two weeks to learn about Chile's problematic fisheries.

The Ley Longueira is legislation that established individual fishing quotas (IFQs) in Chile, privatizing their fisheries. In this system, a set percentage of the allowable harvest of a fish stock is granted to an entity (person, company, ship, or cooperative) who can then harvest, lease, or sell it just like property. The idea behind the law was to end a free-for-all fishing derby in the sea, sometimes called the "fish Olympics," and to halt overfishing in Chile; but there have been conflicting viewpoints about the effect of IFQs on the health of the country's fisheries and on the welfare of its fishermen. Eco Océanos had posted several articles on its website purporting to expose corruption in the legislative process of establishing Chile's IFQs.

I was traveling with my wife, Mónica. Since she was born in Chile and has family there, I'd been to the country many times to visit friends and relatives. Isolated by its geography, Chile is a leggy island hiding within a continent and shaped remarkably like a skinny dried chili pepper. Despite the appearance, the name probably comes from the Mapuche Indian word for "where the land ends." Chile is a hospitable place surrounded by an inhospitable landscape. At the country's northern boundary, the Atacama is the world's driest desert. The rugged Andes Mountains rise to the east. At the southern limit, treacherous waters heave themselves at Cape Horn. The western edge falls into the Pacific Ocean.

European colonists arrived in the sixteenth century, and they quickly subdued the Incan rulers of what is now northern and central Chile. They never could vanquish the ferocious Mapuches in southern Chile. Colonization by Spain was followed by four centuries of aloofness from the rest of the world. Over that time, Chile evolved a unique society. Many of the old ways still persist; it is a country with bountiful natural resources, but where social standing is often determined by heritage. Economic disparity is great.

Although Cárdenas speaks English, Mónica agreed to accompany me to the meeting with him in case I needed translation. I can get by in Spanish, but the dialect spoken in Chile is nearly impossible for most foreigners to understand. Chileans slur their words together; they use a lot of idioms, and pepper their speech with slang—for example, "Estoy pato" (I am a duck) means "I'm broke." I often find myself pondering these expressions, stumbling while the conversation runs well ahead of me.

We strolled the short distance from the subway stop to Cárdenas's office. The streets buzzed with morning activity—sidewalk vendors arranging palettes of flowers, the clanking of dishes from busy coffee shops. Along the way we passed by the government palace, La Moneda. Seeing it called forth black-and-white memories from newsreel highlights of what happened there. In 1973 General Augusto Pinochet's fighter jets strafed and bombed La Moneda to start a coup d'état that overthrew the democratically elected government of Salvador Allende. Afterward, Allende committed suicide in the palace. Pinochet's dictatorship ruled Chile for the next seventeen years. The regime became an international symbol of repression, political terror, and corruption. I learned that the shadow of the coup still lurked over the life of Chileans, including, as we were about to learn, Cárdenas. The palace has been restored, but it's said that a few bullet marks on surrounding buildings and on a statue in the plaza were left as open wounds so people wouldn't forget.

In his office, Cárdenas talked with us for three hours, smiling broadly the whole time, speaking sometimes in English, more often in rapid Spanish. I was lucky to have Mónica along. Cárdenas hammered out words with his hands as he outlined the blueprint to attack the Ley Longueira in hopes of repealing it. His eyes burned with enthusiasm. Frankly, he was mesmerizing. He pulled us into his world with the gravitational force of his fervor.

"All of the fish in the water belong to someone, but most belong to four companies owned by seven powerful families," Cárdenas said. These seven fishing families include names that are among the most recognizable symbols of privilege in Chile. When he named them and their other enterprises, I realized I had heard of some of them.

Cárdenas counted out the steps in the plan that he and his colleagues hatched to repeal the new fisheries law. He amplified his points with the fingers on his clenched fist. "First"—his index finger wagged—"we associated the law with Pablo Longueira." Longueira was an unpopular politician who engineered the act as minister of the economy. Longueira was a member of the far-right political party, the UDI. Thus, the Fisheries Law of 2012 became the "Ley Longueira." Branding the act was a clever move, similar to the tactic used in the United States when Republicans called the Affordable Care Act "Obamacare."

"Second," Cárdenas said, holding up two fingers, "we named the families controlling Chile's fisheries and the politicians supporting them." Thereby, they exposed corruption, bringing shame to all. Cárdenas said that members of Parliament and fishing company executives were under investigation for accepting bribes in exchange for their votes to pass the Ley Longueira. The leader of the largest fisheries union, who once opposed the law and then reversed course, is also under investigation. One powerful Chilean family that controls almost 10 percent of the world's fish meal is being sued for their interference in the election. Longueira himself has been forced to testify in court about his activities and business interests.

"Third," Cárdenas said, emphasizing with another finger, "we initiated a public campaign against the law, gathering some three hundred thousand signatures. Finally"—he elevated his pinkie finger—"a bill has been introduced in the Parliament to overturn the law."

Cárdenas himself was trained as a veterinarian. As a young professional, he worked with the Mapuche Indians on the island of Chiloé in southern Chile, not only on their animals but on social issues as well. After the 1973 coup d'état, he lost his job. Then he got involved in research on marine mammals and issues of marine biodiversity. He did subversive community organizing against the Pinochet regime. The combined experiences led him to his current activism at the interface of ocean health and social justice.

"I'm somewhere between an anarchist and a Maoist," he said quite candidly. I scribbled his words fervidly in my notepad. I got the sense that his philosophy is deeper than the veneer of a political label. He was passionate about human rights and human dignity. The battle against privatization of fisheries seemed part of a bigger war for Cárdenas. His face glowed as he talked about the need to write a new constitution for Chile.

Cárdenas claimed that neoliberal economists at the Universidad

Católica started planning to convert Chile's economy at least fifteen years before the 1973 coup d'état. After Pinochet overthrew the socialist government of Allende with the help of the CIA and Nixon administration in the United States, an economic plan was put into place.

The Chilean economists, known as the "Chicago Boys," had been trained under Nobel Prize–winning economist Milton Friedman at the University of Chicago. Chile would become a laboratory for a neoliberal experiment where, in theory, by privatizing state-controlled industries and resources they could reduce costs, maximize efficiency, and increase profits. Wealth would trickle down to the people. The neoliberals were opposed to the Keynesian capitalistic concept of economics whereby higher wages, full employment, and generous social welfare programs would create consumer demand and fuel growth.

The Chicago Boys pursued their goal of handing over Chile's public resources to private hands with fanatic zeal. First they privatized publicly held banks, properties, and certain industries. This was followed by privatizing government pensions, water rights, telecommunications, and transportation.

Milton Friedman called it "the miracle of Chile."[1] Some people got very rich from his policy of free-market shock therapy, but many others have questioned the fairness of a system that led to a doubling of poverty rates and one of the greatest disparities of wealth in the world.

Although Pinochet gave up the presidency in 1990 after losing a referendum on his continued leadership, he continued to wield influence as commander in chief of the military until 1998. To this day, his economic policies are pushed forward by the momentum of the political process and the influence of money in the system.

In 2002, the government of Chile effectively privatized 80 percent of the country's marine fisheries, but on a temporary basis. Then, in 2012, the Ley Longueira was passed, permanently granting fish quotas, or ownership, to four companies controlled by seven families. The public's ownership of their fisheries was given away, for free.

Cárdenas had campaigned vigorously against the law, and its passage was a defeat for him. When I asked him how that affected him, he pointed out with a wan smile, "After you've experienced the repression under the Pinochet regime, every other setback seems minor and surmountable."

There are various estimates of what percentage of Chile's fish is in direct or indirect control of the industrial families. Cárdenas says it's 92 percent. Others say it's closer to 50 percent. But a huge share of the catch eventually ends up in the families' hands. Part of the quota of fish is reserved for fishermen labeled by the government as "artisanal." These actually are middle-sized seiners who sell most of their catch to the big fish companies to be reduced to meal. The companies loan these fishermen money to buy their boats, and the debt is paid back in fish at a low price, which the companies sell for a large profit. Some of the boats that the government classifies as "artisanal" are owned by non-fishermen. I find it hard to accept the characterization of this component of Chile's fishery as "artisanal."

According to Cárdenas, the small percentage of the fisheries set aside for Chile's ninety-six thousand small-boat artisanal fishermen (and thirteen thousand boats) is too small. Given that these fishermen's families are involved in their enterprise, the number of Chileans depending on artisanal fisheries is more like a half million.

"The goal of the government and that of the fishing industry is to eliminate the artisanal fisheries," Cárdenas said.

Ninety percent of the fisheries' catch of Chile is within forty miles of the coast, and all but 10 percent of Chile's harvest is reduced to fish meal, a dried fish flour. The fish flour used to be spread on crops as fertilizer, but now it's mostly reprocessed and fed to chickens, hogs, pets, and salmon. Chile ranks number two in the world for tonnage of its industrial fisheries. Cárdenas described how the powerful families that control the industrial fisheries have greater interests in mining and development than in fisheries. He said that they use the harvest of marine resources as a quick source of cash to fund their other ventures, kind of like a natural bank vault.

In spite of Chile's long coastline and abundant fisheries, the per capita consumption of fish by Chileans is very low, only about seven kilograms per year. By contrast, Americans—not known as a big fish-eating population—annually eat twenty-two kilograms per person.

Icelanders eat ninety kilograms. But Chile's people used to eat more fish.

Cárdenas suggested that a shift in diet of Chileans was influenced by the Chicago Boys, who engineered a plan to export more of the harvest, thereby bringing in more foreign capital. As a result, local fish was less available to Chileans. Most of the country's fish harvest now is exported. One journalist recently described observing a school in Chile where the students ate canned fish from Asia for lunch. They were within eyesight of a large fish-processing plant. Cárdenas said that 87 percent of Chile's economy is based on exported raw natural resources, all benefiting corporations or powerful private families. The industrial fishery alone is worth about $8 billion.[2]

Although the IFQ system is supposed to be a tool for conservation of fisheries resources—and is advertised by neoliberals worldwide as a hallmark of success—in reality, it has failed as a conservation measure in Chile. In 2016, Chile's government announced in its annual report on the status of fisheries that a devastating 72 percent were in an overexploited or depleted state. This number is up from 48 percent three years before. Important stocks of hake, congrio, sardine, and anchovy have collapsed.

An IFQ system has been in place in Chile for fourteen years, and there are several ways to measure its success: economic, conservation, social, and consumer-based evaluations. Thus far, by many measures, the regulation of the fisheries has been disastrous for both the fish populations and fishing communities. The collapse of hake, whose population plummeted 90 percent since 2001, hit artisanal fishermen particularly hard. Five thousand artisanal fishermen lost their livelihood. From the consumer's viewpoint, good fish are hard to find in the markets and are more expensive.

On the other side, the industry and some scientists blame Chile's woes on the artisanal fishermen for underreporting their catches. As fish populations have declined, artisanal fishermen allegedly harvest more than the allowable catch. They do so in order to make a living, but thereby create a downward spiral of both fish stocks and future income.

Near the end of our meeting, Cárdenas pressed his palms together as if in prayer and said that he is hopeful that the law will be overturned. New fisheries policies are needed to protect biodiversity and ecosystem health while promoting the people's right to eat, have decent work, and protect coastal communities.

A Coastal Excursion

The next day after our visit with Cárdenas, we headed for the coast of central Chile. First, we visited Zapallar, a seaside village about eighty kilometers northwest of Santiago where we'd been several times in the past. Here, a little outdoor café sits at the edge of the ocean on a cape that juts into the sea. Sometimes the waves crash over the breakwater and force diners to retreat for shelter. In the presence of a stiff breeze and the sea air, one feels as if eating on the deck of a sailboat. Small, sturdy wooden boats chug up to a ramp at the café and deliver their catch directly to the kitchen under the watchful eye of stern pelicans that line the jetty. The boats are painted bright white with blue, yellow, and red trim. The café El Chiringuito serves the local sea fare of centolla, machas, locas, corvina, congrio, and excellent pisco sours.

The next day we drove down the coast for about an hour to Valparaíso to meet with Professor Patricio Arana of the faculty of Marine Sciences at the Pontifical Catholic University of Chile. In 2015, Arana won the Premier Prize of Chile's Congress of Marine Science. He's published three books and many scientific papers. Arana has taught at the university for fifty years. His office was immaculate, smelling faintly of musty books and formaldehyde. Sitting behind a polished desk, he was neatly groomed in sweater and pressed shirt. His close-cropped gray hair was carefully combed, and he spoke in measured tones.

If Cárdenas was a wild-eyed idealist, Professor Arana was his antithesis. Arana said that only some species are managed by setting harvest quotas. For species with a harvest quota, artisanal fishermen operate under an open system without individual quotas, but with a sector-wide catch limit. According to Arana, they are supposed to register as fishermen with the government. He noted that there are a fixed number of licenses available, but the government allows more and more licenses to be granted under pressure from the artisanal fishermen. He said that now there are too many fishermen for the resource, and as a consequence, the fish and shellfish populations are declining.

From the mixture of reasoning and opinion blended in his talk, I began to catch a drift of Arana's view. If I were a professional journalist, I might have asked him about his political affiliation right then. (I was trained as a scientist and have thought that science and politics should be separate jurisdictions. In retrospect, I realize that these two endeavors overlap in the field of fisheries.) However, people in Chile like to express their political beliefs. In this case, I just had to sit tight and be patient. I didn't have to wait long.

"This government [in 2016, the Socialist Party of Chile] just gives and gives," Arana said. "Their goal is to force the industrial fisheries out."

Arana explained why it's so hard to manage Chile's artisanal fisheries. He said that there are hundreds of caletas (fishermen's wharves or terminals) along Chile's five thousand kilometers of coastline. When you include the islands and bays, the coast is eighty thousand kilometers. That's a lot of coast to monitor—too much for the Chilean government.

Fisheries in Chile are a complicated enterprise. Because the caletas are far from the marketplace, there are intermediary buyers on each beach. They are like a cartel and protect their beach from outsiders. Another issue is that the artisanal fishermen are supposed to pay a government tax on the fish they catch. Arana said that the artisanal fishermen often don't report catches in order to avoid paying the tax. He reckons that as a result, their actual catch is probably double that of the reported catch.

"The rational view is that the law [of Longueira] is good for the conservation of the fisheries resources," said Arana.

This was his view as a scientist. His goal was conservation of the resource. Arana said he believes that the communists and socialists are trying to remove the law to get rid of the industrialists.

"Besides," Arana said, "the big companies have been fishing in

Chile for a long time. They have historical rights. Why do they have to lose those rights?"

I asked about the discovery and investigation of bribes paid to politicians to support the law. Arana said, "This is just an opportunity [the opposition to the law uses] to say the law is bad."

Arana later described how he was involved in developing the industrialization of Antarctic krill. Finally, he mentioned how good the salmon farms are for Chile. Behind Norway, Chile is the second-largest producer of farmed salmon in the world. Aquaculture in Chile is a $5 billion industry. The 2,300 salmon farms support 120,000 people.[3]

At the very moment we talked, there was a massive red tide—a bloom of toxic algae—developing in southern Chile. Eight thousand tonnes of sardines had washed up on the beach, and forty thousand tonnes of dead farmed salmon were dumped into the sea. Artisanal shellfish harvests were halted.

A few days later, thousands of artisanal fishermen held anti-salmonera (salmon farm) demonstrations on the island of Chiloé, about five hundred miles south of Santiago. The island has many artisanal and indigenous fishermen.[4] They blocked roads with burning tires. The demonstrators in Chiloé blamed the salmoneras for polluting the waters and causing the bloom of algae. This red-tide event was reported as being of biblical proportions. The salmon farmers and government scientists said it resulted from an "El Niño Godzilla."

After our discussion with Professor Arana, we returned to our rental car outside and discovered that a prowler had broken the rear passenger window. We'd parked in a dilapidated neighborhood of Valparaíso next to a disheveled café and across the street from the caleta El Membrillo. The thief took only a couple of jackets and a small kit of medications and cosmetics that had been lying on the back seat. We'd never been victimized by crime in Chile before, and the experience was disillusioning. At this point in our journey, we entered the gray mist of legal fine print in our rental insurance agreement and boarded the merry-go-round of the collision-damage waiver option. Our negotiations with agents of the rental company involved an ever-changing delineation of what the insurance would cover, and whether or not the broken window was an accessory item. Dealing with the

Figure 12. Fisherman Miguel Troncoso Olivares with his boat *Chocolito*, Maitencillo, Valparaíso, Chile. Photographer: K. Bailey.

bureaucracy was maddening. The whole experience was like a metaphor for Chile's fisheries: expectation, disappointment, broken, and robbed; hopelessly complicated by words.

The Caleta

Our next meeting was with a local fisherman, Miguel Troncoso Olivares. We contacted Troncoso through Mónica's sister, who sometimes buys fish from him. Troncoso looked like an earthy and less-brassy version of Tony Bennett.

We found him at his wife's craft booth on the edge of the ocean and across a dirt parking lot from the local fish market. Troncoso said he was six years old when he started crying to his grandfather to take him fishing. By the time he was ten, he was rowing a boat while his father harpooned a type of flatfish called linguado from the bow. Now at the age of fifty-seven, Troncoso fishes and also serves as an official in the caleta of Maitencillo, a beach town popular with surfers just north of Valparaíso.

When I first visited Maitencillo as a tourist twenty-five years before, it was a fishing village with a solo run-down café and maybe a tiny hotel. Small houses lined the unpaved road along the coast. I remember a group of people camped for the summer in multicolored tents at the end of the road; my impression was that they were Romany.

Troncoso said there used to be four hundred people in Maitencillo; now there are thousands. The beachfront is plastered with condominiums, cafés, and tourist shops. Just on top of the cliffs sits the luxury resort of Marbella, a huge development for the affluent residents of Santiago and tourists. This is where we were staying, in the apartment of a relative.

At the beginning of my conversation with Troncoso (Mónica was translating), he smiled broadly and gesticulated with generosity; but he appeared cautious about answering my questions before he could see where I was coming from. It crossed my mind that perhaps his guarded nature was a remnant of the Pinochet era, when people would disappear for saying the wrong thing.

Once Troncoso did loosen up, he expressed yet another view of Chile's fisheries. His wife hovered on the edge of our conversation, jovially correcting him now and then, adding to his comments, and ribbing him good-naturedly.

"Why are you speaking only of yourself?" she said. "Why not the other fishermen?"

I asked Troncoso about how they managed the bay.

"There are robbers from the other caletas that come by night and steal our fish and shellfish," he said.

I made a mental note that the caleta El Membrillo, across the street from where our car was robbed, was an hour down the coast. Troncoso described how the local fishermen take turns monitoring the bay at night. Out of forty fishermen, five watch the bay for outsiders, and if they see an intruder, they contact the police.

Many of the caletas, including Maitencillo, are so-called TURF (territorial-use rights fisheries). As opposed to the IFQs, this type of management system has been fairly successful in coastal Chile; fishermen of individual fisheries co-manage the resource with the central fisheries authority of the government.

Troncoso expressed pride in how his community manages the fishery. He said that officials in the government's fisheries agency tell them how many fish and shellfish they should be harvesting, but they always catch less in order to conserve the resources. According to Troncoso, all the fish and shellfish are sold in the local market.

I asked Troncoso about their method of harvesting fish and shellfish. His gaze turned toward the sea. "We used to free dive in the cold sea without a wet suit," he said. He explained that now they fish for shellfish by diving to a maximum of twenty meters, using wet suits and air compressors. The diver searches for shellfish on the bottom, putting his catch into a basket. When he has enough, he pulls on the line; the basket is hauled to the surface. Topside, they carefully sort and count the catch, and send back the individuals that are too small. "We have a lot of shellfish because we take care of the resource," he said. When they go out for finfish, the fishermen use a variety of set nets with different mesh sizes to match the fish they are targeting.

When I brought up the Ley Longueira, Troncoso reminisced about his early experiences. He said that when he was twenty years old he left Maitencillo and worked on big boats out of San Antonio. They fished swordfish until the fish were all gone. Then he worked on a trawler. He said that the wastage of trawling fish from the sea bottom made him weep. One time they had to dump a load of merluza (hake) that were all juveniles—too small. The ocean was white with the carcasses out to a distance of three hundred meters around the ship.

"It broke my heart," he said. "The big families are taking everything. I saw it. They don't care about the resource. They just want to catch it all."

Troncoso wanted the law changed to better account for artisanal fisheries. He agreed with Cárdenas that the industrial fisheries are trying to get rid of the artisanal fishermen. He said the big players in the industry appear to be controlled by the law, but there's a lot of fudging the numbers, and the government is trying to eliminate the small artisanal fishermen because they don't pay taxes. Troncoso was against the law, but he was afraid that if they repeal it outright, the industrial fisheries will "go wild and take it all." So, to him, changing the law for the benefit of the artisanal fishermen is better.

Now I had a broad range of opinions. Cárdenas wanted to expunge the Ley Longueira, Arana wanted to embrace it, and Troncoso wanted to change it. They all looked like reasonable opinions from their own perspectives.

I asked Troncoso, "Are you happy with your life as a fisherman?"

"This is my life—all of it," he said, sweeping his arm back toward the sea.

"Yes, it's true. More important than me," his wife chimed in with a grin. From the corner of my eye, I saw Mónica and Señora Troncoso exchange a knowing look.

Reflection on a Throbbing Ocean

At night we returned to the condominium on the cliffs above Maitencillo. Mónica's brother and sister-in-law drove from Santiago to join us for dinner. Along the way, they picked up a big slab of reineta (aka pomfret) and some scallops from the market. I played chef.

The condo's kitchen was sparsely stocked, so I was challenged to prepare the meal. I cut the fish into chunks and pan-fried them in olive oil and garlic. I plopped in the scallops at the last minute so they were plump and succulent, then squeezed juice of an orange over the mix and garnished it with some fresh mint from the garden. A stiff breeze lifted from the ocean, so we bundled in sweaters to dine on the patio overlooking the bay. We opened a bottle of red wine and watched the sun sink into gray fog on the horizon.

While the others held a rapid-flowing dialogue in Chilean Spanish—full of those ponderous idioms—my mind drifted. I watched the rolling waves from some faraway storm and thought about the illusion of waves looking as if the water is rushing toward shore and crashing on the beach. But really, the water is just moving up and down. Mechanical energy moves along the wave, pushing the water up; gravity pulls it back down. It's like shaking the loose end of a rope up and down when the other end is tied to a doorknob. It's not like it seems. Our eyes deceive us.

Many other things in life aren't like they seem. An article recently published in a prestigious science journal proclaimed how rights-based fisheries were good for conserving fish populations in countries like Chile. The headline was splashed across several popular science digests. What I found in Chile is a more complicated situation. In the large offshore fisheries, there is a failure of the IFQ system and a deep clash of ideologies. It's not like the article said. On the other hand, the small community-based nearshore fisheries seem more successful. But lumping these two types of fisheries together under the banner of "rights-based fisheries" and proclaiming their universal success is like mixing tomatoes and apples together and saying that they all taste the same because they're red. It's not that simple.

Regulations, limited licensing, and catch limits can accomplish the same goals of eliminating overfishing as IFQs. It's the aspect of "ownership," of buying, selling, and inheriting the quotas that troubles me: non-fishermen owning rights to the fish. To me, there is a difference between making a profit and making a living from the sea. The problem with IFQs is the ideology of private property rights standing in place of regulation, the assumption of fairness in distribution of IFQs, and the assumption that all owners with shares will behave responsibly.

Peeling back the layers of time, one might perceive that Chile's fisheries problems began long ago in the conflict of early European colonists with the indigenous people of Chile over resources. The clash of the Europeans with the Yámana Indians[5] of Tierra del Fuego provides a window to Chile's development. It's a tragic history. The story might seem like a diversion from the theme of small-scale fisheries, but I think it's relevant to many problems, and not only in Chile.

Long ago, the Yámana were the southernmost people in the world. They lived and fished in the archipelago west of Tierra del Fuego, around Beagle Channel, and Cape Horn where the two great oceans collide. Four distinct indigenous tribes lived on or around Tierra del Fuego, each with their own unique lifestyle and language. The

Yámana were the people most closely associated with the sea. They were water nomads in canoes.

Almost all the early European descriptions of the Yámana tell of a people with sturdy bodies, muscular chests and arms, and disproportionately thin legs. In the eyes of the Europeans, the Fuegian Indians' hair was unkempt, their features coarse; they held a perpetually perplexed look on their faces. The Yámana were indifferent to the extreme weather of their land and lived nearly naked. But the warmth of fire was critical to them, and they kept fires burning on turf mats in their bark canoes. They covered their bodies in dark seal grease, and decorated themselves with white clay and red minerals.

Typically, the Yámana lived and traveled in family groups. They paddled from place to place within the archipelago, rarely staying in any one camp for more than a few days. The exception was when a whale carcass became stranded on the beach. Then, the family groups would gather around the source of fat and gorge for days or weeks, providing a social occasion.

The Yámana paddled lightweight bark canoes and dugouts. Their weapons and methods of hunting included bow and arrow, harpoons, spears, and slings. They fished with hookless lines, and pried mussels with forks. The women swam naked in the icy water, diving for shellfish. On land, they gathered berries, fungi, and birds' eggs. Small schooling fish were harvested with baskets. The men hunted penguins with spears, snared geese, and stalked seals and sea lions with harpoons. Otters were hunted with dogs and spears.

I was shocked to read descriptions of the Fuegians written by Charles Darwin, one of the mentors who sit around the fire circle of my imagination. When he encountered the Fuegians, he said, "Without exception [it was] the most curious and interesting spectacle I ever beheld: I could not have believed how wide was the difference between savage and civilized man: it is greater than between a wild and domesticated animal, in as much as in man there is a greater power of improvement."[6] Darwin, although willing to break scientific barriers with the theory of evolution, had a more difficult time breaking from the cultural and religious beliefs he was raised with.

The German anthropologist Martin Gusinde, who studied Tierra del Fuego from 1918 to 1924, had another perspective. He wrote, "The entire landscape is wrapped in a weird gloom," and continued: "A disagreeable dampness soaks everything." His views of the indigenous people differed from Darwin's. Instead, Gusinde said, "For the human beings destined to live in this cheerless region, narrow limits have been set for their demands in life and for the development of material possessions." And later: "Actually the Yámana have not merely adapted themselves to the natural conditions of their homeland to the extent they were forced to do so; mastering their environment, they have created for themselves a dignified human existence of the most modest kind."[7] While not painting a naturalist picture of the noble savage, he said that they were intelligent in a way to adapt to a harsh environment and survive.

Although the mixture of seal grease and fire in a bark canoe doesn't seem like a good recipe for survival, the Yámana prevailed around Tierra del Fuego for at least ten thousand years before the Europeans arrived. Then they perished.

In the mid-nineteenth century there were an estimated three thousand Yámana. In the late nineteenth century an avalanche of colonists descended upon them. Gold miners rushed in to dig for riches. The Europeans slaughtered seals for pelts, and the guanaco was hunted to make room for sheep. The seals had provided the treasured fat that kept the Yámana alive in the harsh climate; seals and guanacos were the life-enabling foods of the Yámana and the other Fuegian tribes.

The colonists took and took, until there wasn't enough to sustain the indigenous people. The Europeans brought new diseases, which ravaged the vulnerable natives. Venereal disease, measles, influenza, and smallpox were lethal to the Fuegians. Missionaries tried to use the Yámana for their own purposes, but these hardy people weren't easily softened and tamed. Over time the Yámana people just became inconvenient. In fear of the natives stealing their sheep, ranchers put bounties on the heads of the Fuegians. Some were shipped off to Europe for display in circus- and zoo-like venues.

By 1908 there were 170 Yámana, and in 1947 only 43 remained.

Finally, in 2016 only a single full-blooded Yámana still lives, an eighty-five-year-old woman. When she dies, there will be none left.[8]

Chile is a beautiful and modern country, but one wonders if the European colonists have simply been replaced by wealthy industrialists and privileged families. "The miracle of Chile," as Friedman called it, left Chile with the greatest inequality gap of the thirty-four nations in the Organisation for Economic Co-operation and Development.[9] A Credit Suisse report says that 42 percent of the country's wealth is concentrated in the hands of the richest 1 percent.[10] Professor Luis Miguel Rodrigo of the Universidad Católica del Norte said, "The ethno-racial dimension of societies with colonial pasts explains why those societies end up having more unequal social groups in a more entrenched system."[11] Not unlike the old colonial powers, who used religious conversion of indigenous peoples to mask their true motives of acquisition of territory and resources,[12] could it be the executives of the privileged fishing companies of Chile and their political supporters veil the motives of corporate control and profit behind a gauze of conservation? Meanwhile, Chile's fish and shellfish populations have diminished. Once again, the indigenous theme of "take no more than you need" has been overrun by the industrial mantra of "take as much as you can." Big fishing companies blame artisanal fishermen for the decline of Chile's fisheries. Artisanal fishermen blame big companies. Some scientists take a more neutral stance and blame the fisheries declines on the weather.

The distant ocean around Tierra del Fuego is a harsh environment where the fish are reported to be abundant still and the waters aren't yet polluted. After I returned home to Seattle, I read Pablo Neruda's poem "Ode to the Book," originally published in 1954.[13] I paused at the lines

> Among the islands
> our ocean
> throbs with fish,

touches the feet, the thighs,
the chalk ribs
of my country.

I reflected back to my conversation with Juan Carlos Cárdenas. There was a point when he told me about the industry's new plans for expansion in the extreme south of Chile, in Magallanes. His eyes on fire, Cárdenas drew a balloon in the air with his finger as he described how the ancestral home of the Yámana, Neruda's islands, are about to be trawled by the industrial fisheries and planted with vast new farms to grow Atlantic salmon in the Pacific Ocean. Then Cárdenas poked his finger through the middle of the circle, popping it—as if extinguishing a dream.

Mermaids

Like the women shellfish divers of the Yámana who swam in the cold waters off Tierra del Fuego to feed their families, the haenyeo, or "sea women," of the island of Jeju off the southern coast of South Korea dive for abalone, octopus, and conch four seasons a year.

The haenyeo used to wear cotton suits, but now they wear full-body wet suits and face masks in the chilly water. They work long hours and dive in waters as deep as 12 meters nearly one hundred times a day, using their hands and spades, or sometimes spears, to collect seafood.

The tradition of women divers began centuries ago when officials demanded dried abalone as taxation. The tradition has continued, but now it's to earn money instead.

In this culture, men used to pay brides a price to marry them instead of the women paying a dowry. "Diving was the lifeline for the entire family," said one female diver. "Men are lazy," she continued. "They can't dive. They are weak under the sea, where it's really life or death."[14] It's said that the men of the Yámana also couldn't swim.

The work is dangerous. Between 2009 and 2014, forty divers died. In the 1960s there were 26,000 divers, but now there are fewer than 4,500, and 84 percent of those are sixty years of age and older. Now the younger women prefer to work in the tourism industry that thrives on the island.

...

"They carried the joys and sorrows of those living with the sea," wrote photographer Iwase Yoshiyuki of the ama divers of Japan. The tradition of ama divers dates back two thousand years. The divers used to wear nothing except a loincloth but now wear wet suits or loose-fitting white linen on their dives searching for abalone, seaweed, and pearl oysters. They stay underwater for up to four minutes, making sixty dives each session. These days many of the divers work for the Mikimoto pearl company, tending oyster farms.

In 1956 there were 17,611 ama divers in Japan. But the profession became less attractive as other career opportunities opened to young women. By 2010, only 2,174 divers remained.[15]

There is an ama saying: "When a man comes to the ocean, he exploits it and strips it. When a woman puts her hand in the ocean, balance is restored."

LOSS AND RECOVERY OF INDIGENOUS FISHERIES

They always spoke about the long, long time ago—a time so far, so far gone that we are looking at it through a mist or a fog. It is our time when we were on this earth. We have been here for a long, long time. We have always been here.
Harriette Shelton Dover, Tulalip, from My Heart

Stories of indigenous fisheries track the beginning of humanity's relationship with aquatic resources. Fishes have adapted to a changing landscape over a very long time. Likewise, indigenous cultures that harvested fish were closely attuned to the ecosystems around them and were—and still are—characterized by remarkable flexibility. Over the past several centuries, when the lands of the indigenous peoples were colonized, their relationship with fisheries changed from subsistence to commercial activity as they engaged in trade with the newcomers. Over varying scales of time, many indigenous cultures lost their fishing rights. In the past several decades, there has been a struggle to regain those rights; some of the battles have been won, while others are ongoing. Indigenous peoples have been among the most active groups involved in raising awareness of threats to fisheries, especially in the area of habitat degradation.

I believe that to understand what has happened in fisheries, one has to understand what happened to the fishing people. Out of many possibilities, I've chosen to tell the stories of two indigenous cultures, the Coast Salish people of Puget Sound (chap. 4) and the Sámi of northern Norway (chap. 5), because of their well-documented histories and their continuing struggles for fishing rights.

4: The First Fish

THE COAST SALISH SALMON FISHERY

In June 2015, I watched a band of people, many wearing red tops or woven cone-shaped hats, march alongside the Snohomish River in Washington State. They followed two young men who carried a salmon, resting atop a bed of ferns, between them. To the beat of drums they chanted, "O choba tubo—yo bouch teah."[1] The fish was the symbolic first-caught king salmon of the season. As this guest entered the ceremonial cedar longhouse, a group of Tulalip tribal members waiting outside the door raised their arms to greet it.

Inside, spectators filled the elevated benches that lined the sides of the building. They watched a procession of fifty men parade behind the salmon, followed by women and children. Rhythmic chanting, thunderous drumming, and swirling colors of dancers merged with smoke billowing from two great fires in the filtered light. My eyes burned and filled with tears.

The salmon was cooked and passed around the room. Each person took a small portion, sharing with the community. When the tribal members were done, they returned the salmon's bones to the sea.

I make my home near the shores where Native American fishermen thrived for thousands of years, and where their descendants still live. According to their stories they have been here not just for years, but since the beginning. Watching the ceremony felt like receiving a gift left behind by the ancestral people of the Salish Sea.

The lives of the fishermen and that of all the Salish people began to change with the arrival of colonists. It's a story that's repeated over and over in the history of every continent. What happened to the salmon people is a conversation we need to have. In the words of the

Figure 13. The arrival of a canoe carrying the symbolic first fish during the Tulalip First Salmon ceremony, June 2015. Photographer: K. Bailey.

fisherman, Native American activist, and Nisqually tribal elder Billy Frank Jr. when interviewed by Charles Wilkinson: "Tell them about our life. Put out the story of our lives, and how we live with the land, and how they're our neighbors. And how you have to respect your neighbors and work with your neighbors."[2] The following chronicle tells how our attitudes toward the ecosystem and our views of our place in the natural world have changed. It's important to recognize the multiple perspectives of the indigenous and colonial cultures, the views of tradition and science—and to tell the story with "two-eyed seeing."[3]

From 1833 to 1859, James Murray Yale[4] was the chief trader for the Hudson's Bay Company at their outpost in Fort Langley in Canada. He often traveled from the isolated fort on the Fraser River due east of Vancouver, down into Puget Sound to establish trade relations with the local Indian fishermen and trappers. Whether he knew it or not, Yale played a major role in the struggle between European colonists and native tribes for fishing rights, as well as in a major ecosystem shift along the Pacific coast caused by the fur-trapping business (see chap. 9).

Perhaps on one of his journeys south, Yale visited the Snohomish tribe and, like me, was permitted to watch the First Salmon ceremony. I picture Yale encountering a Snohomish warrior and fisherman named Qaba'xad.[5] It seems probable that they met. Qaba'xad is known to have fished and trapped for the traders at Fort Langley.

As I envision it, Yale might have seen Qaba'xad working as the fisherman spread his fishing mat on the riverbank to prepare for the salmon's return. His calloused fingers stroked over the woven cedar bark mesh to release a woody smell as he searched for rips. Qaba'xad replaced some white shells embedded in the bottom of the net. He told Yale that the shells let him see the dark shapes of salmon swimming over them like silhouettes across the moon.

The river beside them ran swift and deep from the snowmelt in the mountains. There was something in the scent of the river that awakened a memory in the salmon that had been born there; it enticed the salmon to return from the ocean to their birthplace. Every year at this time the fish ran up the Snohomish River. When they got to the confluence of the Skykomish and Snoqualmie forks, which unite to form the great stream below, instinct drew the salmon deeper into the mountain's shadows. Here they would spawn, die, and decompose in the cycle of life that would sustain future salmon runs.

Yale observed that after a winter of eating dried fish, the people of the Snohomish waited with anticipation for the salmon to return. Qaba'xad said that when he saw a black-and-white butterfly named Yuyubaĉ ("the little king salmon"),[6] he would hurry back to the village to tell his people the time had come to leave for the summer camp and begin fishing.

Then the men and women from all the longhouses got in their river canoes and traveled down to the mouth of the river where it ran into the salt water. From here Qaba'xad and his family group got into their big sea canoe, and the others did the same. They paddled about forty miles to the islands where they made their summer camp. The fishermen would troll the waters with shell lures tied to braided kelp lines—trailed by bone and myrtle hooks—to catch the first king salmon returning to this place. Later they would place the nets for the big runs of the sockeye salmon.[7]

The fisherman told Yale that the salmon always came back if they were treated with respect. They left as small fish and reappeared as adults. The elders of the tribe said the salmon were a race of beings that lived in a village under the sea where they took human form. Every year some of them donned silver skins and swam to the rivers as a gift of the salmon king to the coast people. Often the first salmon to show up were the biggest fish, maybe because they were more powerful swimmers, or maybe their urge to spawn was greater than that of the younger fish.

Perhaps Yale observed Qaba'xad as he hooked a monster, the first fish of the season. It strained the braided line. After he wrestled it from the water, Qaba'xad cradled the fish in his arms and spoke to it with reverence: "Welcome, noble one."[8] This fish was taken back to the fishing camp for sharing.

The fishermen returned to the camp for a great ceremony.[9] The people treated the first fish with reverence. The salmon was placed on a bed of ferns. The people took care not to step over the fish, a sign of disrespect. The women could not step in front of the fish. Qaba'xad's wife filleted and roasted the salmon on ironwood skewers over coals.

When the feast was over, the salmon's bones were placed on a mat. Qaba'xad took the bones back to the place the fish was caught and released them into the water, facing toward the west. He told Yale that the salmon's spirit would tell the other fish that he was treated with respect by the men; then they knew they would also be treated with respect, and they would follow to the river in plenty. After dying, their spirits would cross to the other side. The people of the tribe believed that the spirits of these fish, in turn, would feed their own ancestors in the spirit world.[10]

The women made nets and ropes from cedar roots that grew thirty or forty feet down from rotting nurse logs that hosted the sprouting trees.[11] They also made lines from dried nettles braided together with the inner bark of cedar or willow; as well as from kelp stipe that had been soaked in freshwater, stretched, twisted, and spliced together. Stones wrapped in cedar or cherry bark, or drilled with holes, were used as sinkers. Trolling was done with shell lures or with steam-bent hemlock for hooks. The hooks were barbed with a bone spur and baited with clam meat or a small fish. Almost every imaginable method was used to catch the salmon. Qaba'xad built weirs along his tribe's land on the Snohomish River. He used the weirs to trap fish, and then he netted or speared the fish in them. He also deployed gillnets or beach seines, depending on the species he was hunting, the area, and the season.

The marine waters were neutral grounds for all fishermen, and although nearshore areas beside their villages were tribal territories, they often invited other tribes to fish the area with them. Intertribal marriages cemented linkages among them. They fished the Chinook, sockeye, pink, and silver salmon runs around the islands offshore (the San Juans) and in the Snohomish and Stillaguamish Rivers. The fishermen traveled north to catch the big runs of sockeye salmon returning to the Fraser River and finally ended with the autumn runs of dog salmon in Hood Canal.

Yale knew that when the fishermen traveled, boats from different tribes sometimes traveled together, the Snohomish along with the Swinomish, Lummi, Skagit, Suquamish, and others. Often there would be thirty or forty canoes. They banded together for protection from the people of the north who spoke a different tongue, the Laich-Kwil-Tach and other bands of the Kwakwaka'wakw group. These northerners were fierce people who often raided the Coast Salish, pillaging their stores and capturing slaves. Sometimes people from even farther north, the Haida and the Tlingit, attacked them as well.

Cedar, Salmon, and Culture

The Coast Salish people and the Pacific salmon flourished beside each other, but according to Western scientific studies, it wasn't always that way. Archaeological data show the first people in the Pacific Northwest arrived twelve thousand to fourteen thousand years ago

Figure 14. Lummi fishermen setting a reefnet between two canoes, Bellingham, Washington, about 1930–33. Photographer: Eugene H. Field. Used with permission from University of Washington Libraries, Special Collections (image NA1813).

at a time when the Cordilleran Ice Sheet was in retreat. The Coast Salish traditional knowledge puts the people in the tribal lands much earlier, and more recent scientific estimates now put the initial arrival of humans on the continent back a full ten thousand years.

Geological evidence shows it must have been a hardscrabble existence for the first people here. They lived on the fringe of the ice, in pockets on the Olympic Peninsula, and along the shoreline. Few traces of these earliest coastal inhabitants exist now because the sea level was about three hundred feet lower in that era and most evidence of them vanished when the ice melted. The rising ocean covered their remains.

Toward the end of the Pleistocene, a giant ice sheet covered Puget Sound and its rivers. Four thousand feet of ice loomed over what is now Seattle. As the climate warmed and the glaciers retreated, they scoured the bedrock and carved out much of Puget Sound. Left behind were the large river basins that drained the hills of winter rains and the mountains of melting snow in summer. The rivers were filled with glacial till, fine sediments that smothered animal life.

The complex ecosystems necessary to sustain young salmon in the rivers weren't in place for thousands of years. Some rivers that existed during the glacial period—the lower Columbia and rivers farther south, and perhaps a few rivers on the Olympic Peninsula—may have served as refugia for small residual salmon populations, but these rivers were periodically swept by colossal floods: disasters for salmon trying to live in them.

Salmon are fishes that readily colonize appropriate habitat. They have survived many ice ages in the Pacific Northwest over time scales of millions of years. They are adaptable to a wide range of environments. But before the salmon could reproduce and thrive in the postglacial rivers of the Pacific Northwest, the plant and animal life needed to sustain them had to take root.

The process of ecological recovery in a raw landscape is called *succession*. In the Pacific Northwest, contemporary studies at the edge of retreating glaciers infer that the postglacial succession process began with growth of a black crust of algae that stabilized silt and retained water. Then moss and lichens attached to the rocks. These were followed by plants like horsetail, fireweed, dryas, and willows. The seeds of big trees—perhaps carried by birds, mammals, or the wind—sprouted and began to set their roots; first came alder, then spruce, hemlock, Douglas fir, and western cedar.

When the composition of the plant community changed, it altered the habitat. The organic content of the soil and the makeup of the nutrients within it shifted, as did the amount of light and shade reaching the forest floor. The microbial community established itself, recycling dead organic matter.

The earliest fir trees were scrubs, barely surviving. The dense Douglas fir forests didn't appear until about four thousand to six thousand years ago when the soil, nutrients, and recycling community of bacteria and fungi took hold. The roots of the ancient fir and cedar forests were needed to stabilize the stream banks and make the rivers optimal habitat for salmon survival. The expansion of the western red cedar forests, which is so closely linked to the woodworking culture of the Coast Salish tribes, took place only three thousand to five thousand years ago.

But other conditions in the salmon's habitat had to develop as well for the fish to thrive in the Pacific Northwest. Most of the salmon's life is spent in the ocean. When the ice melted, the freshwater flowing into the ocean added minerals to the sea and changed the buoyancy of the upper part of the pelagic zone, supporting more productivity. The freshwater and shifting winds also changed the ocean currents. During the glacial maximum, the ocean water was too cold for optimum growth of salmon. About three thousand to four thousand years ago, the climate began to heat up, the northeastern Pacific Ocean warmed, and circulation along the coast changed. Ocean productivity increased, providing an ample food supply for the salmon.

Remnant populations of salmon in the south and in refugia along the coast expanded their range in the northern Pacific as they foraged for food in the ocean. As the rivers and estuaries began to support aquatic life, the salmon in the ocean tested the rivers for spawning success. They were like gamblers rolling the dice to see if their young could survive, and when they did, the salmon colonized the newly available habitat. They spawned in the rivers, and their offspring returned. After many generations, the salmon proliferated.

The local humans prospered along with the salmon and the forests as the coastal environment changed. According to scientists, this didn't happen instantly or synchronously. The coevolution of the Pacific Northwest landscape and the human-salmon economy developed over a period of two thousand to five thousand years in a patchwork of complexity.

One theory about human emergence on the continent deriving from archaeological studies is that a group of the first people to arrive "followed the kelp." That is, about eleven thousand years ago some Indians inhabited the shoreline and migrated down the coast. In both California and British Columbia, there is evidence that the early people there already had sturdy boats and tools. They hunted seal and otter, fished for lingcod, rockfish, and halibut, and they gathered shellfish. Over time, the first people began using bone tools, adzes, gorge hooks, slate knives, and more sophisticated spear points and arrowheads. One imagines there may have been different waves of people moving across from Asia and down the Pacific coast of North America, battling and blending among themselves as they moved across the landscape.

Scientists don't have a complete understanding of the early development of the Coast Salish. There are few archaeological sites in the Puget Sound region older than five thousand years. That doesn't mean they didn't exist before then; the artifacts simply may not have survived the ravages of time. There is evidence that people harvested salmon seven thousand to eight thousand years ago in the Columbia and Fraser Rivers, but the salmon were only locally and seasonally important. Archaeological digs show that six thousand years ago salmon were increasingly important to people on the Bella Coola River in British Columbia; by 3,500 years ago salmon made up 94 percent of recovered animal bones from their trash sites.

In general, somewhere between 2,500 and 3,500 years ago the lifestyle and economy of the coastal Indians began to change from a subsistence-foraging lifestyle to one that amassed, processed, and stored resources during the warm seasons of plenty for use in the cold seasons when food wasn't as readily available. This was after the Douglas fir and alder forests began to dominate the coastal streams, and stands of the western red cedar and spruce had taken hold. The people established permanent winter residences and seasonal camps to take advantage of the shifting salmon runs. They continued to forage for edible roots and berries in the forests and to gather shellfish from the shore. They mastered the use of cedar wood and discovered new technologies that the unique character of this wood allowed.

The Coast Salish tribes took advantage of the salmon and cedar resources to develop an original and complex culture. Tribes like the Snohomish depended on salmon, a resource that replenished itself if it was conserved. They used cedar to make their fishing mats and nets, longhouses, and canoes. These were innovations for catching, processing, and storing the fish. Because of the abundance of preserved salmon, the Coast Salish didn't have to spend all winter foraging for food, and they could spend time on their art and culture.

The tribes prospered as the economy of mass harvest and storage matured. Over time, the tribes and their languages differentiated, and trade developed among the different groups. Harvesting enough salmon to store for the winter also required a critical mass of people to catch and process the fish, which was enhanced by development of community infrastructure and social organization.

Some of the tribes weren't as dependent on salmon, but used a variety of the resources around them. The Samish and Lummi fished for many different species, like halibut and lingcod, and gathered "beach food" like shellfish and herring roe. The inland tribes remained mainly hunters. The Snohomish rarely, or only seasonally, hunted for great land mammals, but they traded for deer and elk meat with the Snoqualmie tribe upriver. They obtained mountain goat wool from the Snoqualmie and Skykomish people in exchange for sturgeon and seal meat, dried salmon, or shell money. The Snohomish also raised a special type of dog for its fur and spun the dog and goat hair together to weave blankets, which they traded, along with slaves and shell money, with the master builders of seagoing canoes: the Makah and other tribes of the outer coast.

The Europeans Arrive

The first Europeans exploring the US northwest coast were surprised to find villages of split-plank cedar longhouses and seasonal camps that were used year after year. They reported heads prominently displayed on pikes at village entrances to intimidate uninvited guests.

The first documented contact of the Coast Salish tribes with outsiders was in the eighteenth century. The English Captain James Cook obtained otter pelts from the Indians in 1778 and sold them at great profit in China. Journals of the Hudson's Bay Company in Fort Langley in the early 1800s describe trade with the Snohomish tribe, who swapped dried fish and pelts of otter, beaver, and seal for iron, copper, cloth, and beads. These were new items the Indians hadn't needed to develop a rich culture in a world of abundant resources, but they were commodities the Indians readily utilized when they became available from the Europeans. The Coast Salish people seemed as adaptable to Western culture as the salmon were to shifting river conditions.

When the Europeans arrived, they couldn't catch the salmon themselves. The fish swam too fast for their heavy boats. Explorers were shocked when they encountered a native fishing technology superior to their own. One of the early explorers in 1855 said, "There is little in the art of fishing we can teach these Indians."[12]

The Native Americans began to catch salmon to feed the Europeans. Every year the Snohomish fishermen paddled north and caught salmon in the Fraser River for the Hudson's Bay Company, and they also brought beaver and otter pelts to trade in the winter. The Hudson's Bay Company salted the salmon in barrels and shipped it to Honolulu and San Francisco. Later they sold fish to the glut of hungry gold miners and settlers who arrived after gold was discovered on the nearby Thompson River in 1857.

Some traders at the Hudson's Bay Company—such as the chief trader Yale at Fort Langley—saw the future in preserving, selling, and exporting salmon. Yale pushed for more and more production. Without the salmon, the newly arrived colonists would have had a hard time surviving. By the 1850s Yale was producing two thousand barrels of salted salmon per year, or about 150,000 fish. Due to the trade of salmon for European goods, the coast tribes unwittingly supported more and more invaders, who would deprive them of their most valuable asset: the land. Yale also demanded more otter pelts from the Indians, especially as this item became scarce and thus more valuable. Yale further arranged trade deals with Russia for otter pelts, and the populations of otter continued their precipitous decline, changing the whole structure of the coastal ecosystem (see chap. 9).

The Coast Salish people adapted readily to European economics, and with the arrival of foreign traders they began catching salmon and fur-bearing mammals in great excess of their own needs. The resources became commodities to trade and sell outside of their traditional society; now they could obtain goods offered by the European traders, which had become symbols of status.

The arrival of the Europeans brought other, more devastating changes to the Coast Salish tribes. Diseases for which the Indians had no immunity—including smallpox, tuberculosis, and influenza—decimated populations. Smallpox epidemics began with the first contact with Europeans in the late 1700s, followed by major epidemics in the 1830s.

In 1858, the Fraser Canyon gold rush swelled the number of non-Indians in British Columbia from six hundred to thirty thousand Anglos in just three months. The demand for protein-rich salmon was enormous. With the Indian populations already depleted by epidemics, non-Indian fishermen began to supply the market for salmon, a role they had not taken before. And as the gold ran out, many miners began to fish the sea.

Another epidemic struck the Native Americans in 1862/63. While diseases diminished the body of tribal populations, the next move, by Governor Isaac Ingalls Stevens of the Washington Territory, wreaked havoc on their spirit.

The Treaties

Governor Stevens began to negotiate treaties with tribes in the region in 1850.[13] His intent was to pool together numerous bands and tribes into concentrated populations living on reservations and to eliminate various tribal land claims in order to free up land for European settlers. Meanwhile, the Donation Land Claim Act of 1850 and an amendment in 1853 allowed the invaders to claim land before a treaty was negotiated and signed. Individuals could claim 320 acres, and a married couple could double that claim; with a shortage of wives, some European men sought to wed Native American women in order to obtain more land through the marriage clause.

In 1855 the Treaty of Point Elliott was signed, and the Tulalip, Lummi, Suquamish, Port Madison, and Swinomish reservations were established. Other treaties (Medicine Creek, Neah Bay, Point No Point, Quinault) followed for other tribes. The Tulalip reservation was not designated for one tribe, because there actually was no single Tulalip tribe. Instead, various tribes were lumped into a confederation on Snohomish tribal lands. The tribes inhabiting the Tulalip reservation were to include the Snohomish, Skykomish, and Snoqualmie as well as some Stillaguamish, Suiattle, Samish, and allied bands.

European settlers wanted unhampered access to the land and its resources. Native American leaders were warned that if the tribes didn't move onto reservations, they would face an onslaught of invaders. Their villages were burned down to force their hand. The representatives of the Snohomish signed a treaty written in English they didn't fully understand. Half of their people had already died of smallpox, and the treaty offered medical care, education, and a measure of protection from the invading settlers by setting up reservations. The tribal representatives signing the Point Elliott treaty gave away millions of acres in exchange for $150,000, a school, a promise of medical care, and the small reservations.[14]

The Treaty of Point Elliott wasn't ratified by Congress until 1859. Although the signature tribes had been promised medical care and resources to start farming, no assistance would arrive until ratification. The intervening years were marked by ravaging diseases and malnutrition. Even though the Tulalip tribes had never farmed in the Western sense, Governor Stevens insisted that they would become farmers. Yet there was no access to roads or water, and the dense forests made poor farmland. There was no money for tools or seeds. The desire to hunt and fish, and the poor quality of the land for farming drove many Coast Salish people back to their ancestral lands, where they clashed with the arriving settlers.

In the 1870s the reservations were placed under direct control of Christian organizations to assimilate them into the European-style cultural tradition. Government agents forced children into boarding schools; the missionaries banned speaking Lushootseed and other indigenous languages, and forbade traditional ceremonies, such as the potlatch.

The Loss of Fishing Tradition

Although the Point Elliott treaty guaranteed protection of traditional fishing grounds and berry-picking and root-gathering sites, it was not honored. As the European settlers encroached on indigenous fishing grounds, they competed with tribal fishermen for salmon. At first there was no intent to interfere with tribal fishing, because it provided much needed food to newly arriving settlers. Isaac Stevens noted in a letter of 1854, "They [the Indians] catch most of our fish, supplying not only our people with clams and oysters, but salmon to those who cure and dry it."[15] But later, rules were made to discourage Indian fishermen and to lessen their competition with non-Indian fishermen.

In 1871 the territorial legislature prohibited most types of Native American fishing gear in lakes. In 1890 the (now) Washington State legislature outlawed salmon fishing in northern Puget Sound from March to May, halting the traditional Indian Chinook fishery. In the 1890s the legislature passed laws restricting Indian fishing to rivers within the reservations. But non-Indian fishermen were allowed to set up fish traps at the mouth of rivers, letting few fish pass and reproduce. The salmon populations began to decline. Most of the salmon that were left got intercepted by the non-Indian fishermen before they reached the streams where the Indians fished.

In 1897 state legislators banned traditional Native American weirs in the rivers of Puget Sound and closed all fishing within three miles of the mouths of tributaries, except for direct subsistence use by the Native American fishermen. Landowners restricted the access of Native Americans to their customary fishing grounds. In 1907 the state legislature closed all Puget Sound tributaries above the tide line to taking of salmon except by hook and line, excluding the reservations (however, there are reports that state game officials also tried to enforce restrictions there[16]). Over the years, more restrictions were piled on. Still, the salmon declined in abundance due to overfishing.

Later, huge dams built on the major rivers further decimated the salmon populations. The situation was reminiscent to Henry David Thoreau's observations on the East Coast, in *A Week on the Concord and Merrimack Rivers*: "Salmon, shad, and alewives were formerly abundant here, and taken in weirs by the Indians, who taught this method to the whites, by whom they were used as food and as manure, until the dam, and afterward the canal at Billerica, and the factories at Lowell, put an end to their migrations hitherward."

The "salmon people" no longer had salmon. Eventually, many tribes like the Snohomish lost their fishing tradition. The Huchoosedah[17]—the Coast Salish people's cultural knowledge and knowledge of self—largely disappeared, or at least was hidden. After the social order and traditions of the tribes were destroyed, there was cultural chaos. Author and researcher Ruth Kirk called what was left of the Makah after the smallpox epidemic of 1853, a "society of broken souls."[18] Cultural losses and diseases probably had a similar result on other tribes of the Pacific Northwest.

Qaba'xad, the fisherman warrior, didn't in fact live to see what would happen to his tribe at the hands of the outsiders. The Coast Salish tribes were frequently warring against their neighbors to the north. Groups of warriors, often from different tribes, banded together to protect their territory, commodities, and winter stores from invaders, or to make retaliatory strikes. In about 1845, a large group of warriors from several of the Coast Salish tribes got tired of a pillaging band of the southern Kwakwaka'wakw (Kwakiutl) called the Laich-Kwil-Tach. The Coast Salish leaders met on Nisqually land to decide who would join in the fight. Warriors from the Nisqually, Suquamish, Dungeness, Skagit, Lummi, Cowlitz, Skykomish, and Snohomish tribes agreed they were ready to battle the northerners.

Qaba'xad said, "I'm going to die or kill Kwakiutl, one of the two. All the time they are raiding us Snohomish, and now I'm mad. I'm going to Kwakiutl."[19]

The Coast Salish men, Qaba'xad among them, assembled an ambush to wait for the invaders near Maple Bay, Canada.[20] They were led by Kitsap, a respected warrior of the Suquamish. Two hundred canoes with ten to twelve men in each of them paddled into position. When the Laich-Kwil-Tach warriors came down from Johnstone

Strait in their canoes, the Coast Salish fighters attacked them with spears, arrows, and clubs. Many men on all sides were killed, but the warriors from the north were vanquished, and their villages, now unprotected, were destroyed. The Coast Salish killed nearly all the Laich-Kwil-Tach and kept some as slaves. Here at the Battle of Maple Bay, Qaba'xad was mortally wounded. His body was returned to the Snohomish lands.

Return to Fishing

It was the Anglos' canneries that got the Snohomish back to fishing. In the 1940s, facing a labor shortage due to the war, the New England Canning Company commissioned some of the Tulalip tribesmen to start a commercial fishery, although they were limited to beach seining. Finally, in the 1960s they were permitted to use gillnets and could now compete with the non-Indian fishermen for an ever-diminishing supply of fish. This began a new era of conflict with non-Indians based on access to the salmon resource. Many of the salmon returning to their natal streams were intercepted offshore by non-Indian trollers, purse seiners, and gillnets. The Indians who waited for the salmon by the rivers were the last in line to catch the remaining fish.

The supply of salmon in Puget Sound dwindled, and not just because of fishing. There was a multitude of factors causing the demise of salmon. Dams were built for irrigation and cheap power. Forests were cut for lumber and to clear land for farms, and development degraded the river habitat. The drainage and waste from mines, ranches, and farms polluted the water, as did runoff from roads in urban areas. Farmers and foresters used pesticides that poisoned fish. Industries dumped toxic waste in waterways, and cities dumped raw sewage. The harvest of salmon in the Puget Sound area decreased by more than half between the end of World War II and 1960, and continued to decline in a free fall. The competition for salmon became fierce.

The inequity of catches set off physical battles between Indians and non-Indians, and legal battles between the state of Washington versus the federal government over treaty rights. The Indians insisted on their rights to fish off the reservation on traditional fishing grounds if the non-Indian fishermen weren't leaving them enough fish to catch.

The state was firmly aligned with non-Indian fishermen against the Indians. State officials broke up Native American demonstrations with tear gas and clubs and cut the gillnets of tribal fishermen both on and off the treaty-protected waters. A Nisqually tribal member named Billy Frank became the face of the movement for Indian fishing rights. He was arrested over fifty times, the first time when he was only fourteen years old. He brought in celebrities to popularize his cause of Indian fishing rights, including Marlon Brando and Dick Gregory. Violence on both sides escalated.

The federal government and treaty tribes sued the state of Washington over the tribes' right to fish outside of reservation boundaries. US District Judge George Hugo Boldt, a conservative Eisenhower appointee to the federal bench, was assigned to the case. Boldt was a law-and-order judge, with a strict sense of decorum. State officials and nontribal fishermen anticipated that Boldt would be sympathetic to their side.

In a trial that lasted four years, the treaty tribes argued that non-treaty fishermen intercepted all the fish before they arrived at the reservation, leaving few for Indian fishermen. By this time, Indian fishermen were catching only 5 percent of the harvestable salmon. The state argued that the treaty rights had been extinguished by time, and what's more, the treaty rights only extended to the Indian fishery on the reservation.

Boldt's long-anticipated decision in 1974 surprised everybody. He ruled that the treaty rights remained in accordance with the original wording in the treaties of 1854–56. The treaties stated that Indians had rights to fish "in usual and accustomed places . . . in common with all citizens." Boldt's understanding of "accustomed places" meant that the Indians could fish off-reservation, as they had historically done. He interpreted "in common" as the rights of the treaty Indians to half the fish.

After the ruling, opponents of the Boldt decision resorted to vigilante violence against Indian fishermen. Some state officials said that the ruling established a "hereditary aristocracy." The state refused to enforce the Boldt ruling, and the non-Indian fishermen openly rebelled against it. A filing by the state to the Court of Appeals was unsuccessful, and they appealed again to the US Supreme Court.

In 1979, four Democratic and five Republican members of the Supreme Court of the United States convened to make a definitive decision about Indian fishing rights. Jimmy Carter was president of the United States. In a bipartisan six-to-three majority opinion of the court, Justice John Paul Stevens wrote, "Both sides have a right, secured by treaty, to take a fair share of the available fish." The Supreme Court's review of the Boldt decision reaffirmed Indian treaty rights and determined that they had a right to one-half the harvestable stocks of migrating salmon, excluding the number needed for spawning escapement, tribal ceremonies, and on-reservation catches.

Since the decision, the treaty tribes have co-managed the salmon stocks in Puget Sound and take half the harvestable salmon. Tribes not originally signing the treaty, or not recognized by the federal government, have no treaty rights. They have to fish for salmon as if they were non-Indians. Salmon harvests are dutifully managed and regulated. Fishing pressure seems to be a minor issue for sustainable salmon stocks compared to the other conditions such as pollution, effects of dams, and loss of habitat.

Cultural Revival

A little more than a century ago, missionaries banned the ceremony of the first salmon. The ancient ritual was lost. Harriette Shelton Dover, in her book, *Tulalip, from My Heart*, told how the Tulalip people regained this living tradition. People searched their memories to bring back fragments of customary practices, and tribal elders pieced them together as best they could to recover the First Salmon ceremony.[21]

Billy Frank Jr. said, "We have ceremonies for the first salmon of each run. We bring everybody together and share the first salmon and we train our children that way. When we eat the salmon we give our offerings to the fish and the river. We're not separate from the river. Indian people don't have a cathedral. We have the land and the river."[22]

For the Tulalip, salmon fishing has returned but remains a relatively small resource in the economy of the tribes and as a source of employment. Things have changed on tribal lands; now casinos dwarf any other employer on the reservation. In 1972, prior to the casinos, the tribal government employed only fifteen people. Most of the young men fished. Fishing was not only a source of income, but a rite of passage and a way to prove one's worth. A family fished together, thereby creating close bonds. Now only sixty to one hundred families make part of or all their income from fishing, while far more tribal members work for the casino or in tribal government supported by the casino revenue.

The Tulalip and other tribes have reinstituted a First Salmon ceremony, but reclaiming the cultural significance of salmon for modern-day Coast Salish people will be challenging (from my viewpoint). Things will never be the way they were. The majority of the Tulalip people are disconnected from the fishing tradition. I asked an elder from another Northwest tribe about the disparity between the traditional and modern ways of life. "Culture is a process. We aren't frozen in time. We can't go backward. We can't get back what we had then, only preserve what we have now," he said.

I learned about a business meeting of the Tulalip, where the tribes' economic advisors questioned the cost-benefit of their salmon hatchery. After all, the profits from the tribes' casino and shopping mall operations are supporting most of the other activities on the reservation. Without this funding, the hatcheries would not be able to operate, and without the hatcheries there wouldn't be enough salmon returning to the rivers to sustain a fishery. The question was asked: "Is the cost of the hatchery operation worth maintaining?" A few strong voices held steady to reap the benefits of modern American culture in order to replenish the salmon and respect a traditional way of life. Two-eyed seeing.

"It's not about money," tribal elder Ray Fryberg said. "It's about our tradition and identity. We are a fishing people."

The 'Namgis: On the Cutting Edge of Closed-Containment Aquaculture

The 'Namgis is a First Nations band of the Kwakwaka'wakw people.[23] They live in the forested landscape near Port McNeill and on islands offshore of the inner coast of Vancouver Island. In the late 1950s, there was a drastic drop in sockeye abundance on their home stream, the Nimpkish River, which they believe was the result of a DDT poisoning event. Along with a shortage of fish in their river, many 'Namgis believe that their fishermen were furthermore crowded and regulated out of the commercial wild salmon fishery.

Tribal elders have since made a decision to involve their people in the salmon fisheries again. According to Chief Debra Hanuse, salmon is the key to transmission of traditional knowledge to the 'Namgis people. Salmon farms in the area have been controversial. Many scientists have been suspicious of the effects of salmon farms on wild salmon populations, but the body of research is still in active debate. Regardless of scientific uncertainty, the tribe wanted to do something positive, to show another way for raising salmon, rather than just be a voice of opposition to ocean ranching.

As a result, the 'Namgis elders and some investors decided to cultivate salmon in a cutting-edge closed-containment system. The 'Namgis partnered with SOS Marine Conservation Foundation and Tides Canada to finance the farm Kuterra, and construct the first hatchery module in a planned complex at a cost of about $9 million Canadian.

The company advertises itself as a green enterprise that doesn't use antibiotics, hormones, or other chemicals to grow their salmon, and the fish and water used in the facility are isolated from wild fish, protecting them from pollutants, exposure to diseases, parasites, and escapees. According to Chief Hanuse, their intent is to demonstrate to marine salmon ranchers that there is another way which is not threatening wild salmon populations or the marine environment.

The facility recycles 95 percent of the water it uses each day. The discharged water is filtered and chlorinated to kill any remaining microorganisms before being released into gravel settling ponds, where it takes nine months to seep into the sea. Solid waste is shipped off-site, where it is mulched with wood waste and sold as compost. The plant uses geothermal energy to heat and cool the water that is used to grow the fish. The systems maintaining the five large tanks, each holding thousands of fish, are as automated as possible. Everything—from the temperature, light levels, and oxygen content of the water, to the feeding of the fish—is carefully monitored.

To break even on operations, the company needs at least two modules, and they are searching for investors to finance the second module. Their plan is to build a total of five modules and a hatchery to produce their own supply of smolts.

To begin rearing, a cohort of forty thousand smolts are quarantined in isolation in a 250-cubic-meter tank for two months to ensure that they are healthy and disease-free. Atlantic salmon is the fish of choice, as it has few problems in culture and is the most desirable fish in the market. Their feed is pellets, 8 percent of which is from wild forage fish, while the rest is plant-based or poultry waste. According to Kuterra, the ratio of food consumption to yield is 1.25. Compared to typical ocean ranching, Kuterra's fish use 30 percent less feed and grow to harvestable size in half the time. For the last week in culture they are fed natural carotenoid pigments extracted from yeast and bacteria to give the meat its red color. They are harvested in one year at five to six kilograms.

The product wholesales at 30 percent over conventionally ranched salmon and is marketed to consumers who put a premium on sustainability.

5: Northern Lights

THE SEA SÁMI FISHERY IN NORWAY

In the darkened room where we sat, Torulf Olsen's eyes glowed like embers when he talked about the fishing rights of the Sámi—the indigenous people of northern Scandinavia.[1] He leaned forward in his chair and told me about the campaign to win back the fishing rights that were taken from the Sámi people. He told me about the battles to prevent a mining company from polluting his fjord's waters. Then his gaze shifted to the window, and he talked about the struggle he saw in the future. The story of the sea Sámi—like that of many indigenous peoples—is one of persecution and conflict with invading colonizers; a loss of respect, land, and fishing rights, followed by the fight to regain them.

Torulf lives high above the Arctic Circle near the northern tip of Norway. He is a sea Sámi. His house is a splash of color on the tundra at the edge of Kvalsund (Whale Passage) on Repparfjord. The hills around him lift seven hundred meters into a thin atmosphere. On top of the hills, there is no sound but that of the autumn wind screaming across tundra and stone, closely hugging the terrain. Finnmark, the land at the extreme tip of northeastern Norway, is a rugged place. The system of fjords along the coast of Finnmark are like long fingers and thumbs of the Barents Sea kneading into the earth. The sea Sámi have lived here for thousands of years.

A gale at sea claimed the life of Torulf's grandfather, a fisherman. Another storm took his father, also a fisherman. As Torulf told the story, one evening when he was twelve years old, his father rowed him in a skiff out into the fjord to secure their fishing boat as a storm approached. Somehow they turned broadside into a wave; the skiff

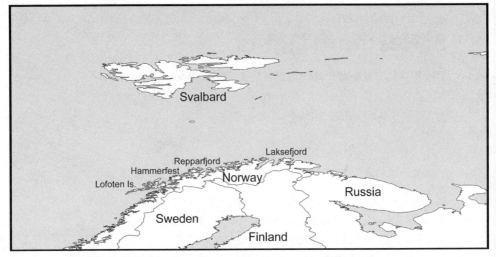

Figure 15. Map of northern Norway showing the area of Finnmark.

capsized, throwing both of them into the water. Torulf never saw his father again. The next day his mother sat him down and said, "Well, Torulf, now it's time for you to take over for your father." Embracing his responsibility, Torulf became a fisherman. Since then he has earned a college degree and had a number of other jobs. Recently he was involved in the business of certifying sustainable fisheries. Now Torulf is the director of Bivdi,[2] a group promoting the rights of Sámi fishermen.

Over the course of centuries the sea Sámi have been denigrated and marginalized in Norway, and they lost their aboriginal fishing rights. But in the past few decades they have been fighting back to improve their standing in Norwegian society, reclaim their fishing rights, and protect their fisheries. The Sámi people have a view of their resources that is different from the modern European, and which needs to be acknowledged as part of the richness of human heritage. The traditional Sámi belief is that they are part of the natural world, similar to other members of the ecosystem, and that nature should be respected, as opposed to the modern European view of nature as something to conquer, tame, and capitalize. The Declaration of the Rights of Indigenous Peoples adopted by the General Assembly of the United Nations in 2008 affirms the rights of indigenous people like the Sámi to the lands and resources that they have "traditionally owned, occupied, or otherwise used" (article 26).[3]

The Sea Sámi in Prehistory

The exact origin of the Sámi is uncertain, but evidence of their existence in what is now Norway dates back to about 8000 to 6000 BC. They came during the Stone Age migrations of northern European hunter-gatherers, during a time when coastal Scandinavia was free of ice. Probably they intermingled with Asiatic people who migrated west along the northern Eurasian shoreline.

Over many centuries, an encroaching Norwegian civilization eventually overwhelmed the Sámi culture, and the story of the indigenous people became a footnote in the history of the dominant society. Much of the Sámi history is lost. Early European maps of Sámi lands marked their territory as "uninhabited."

The early sea Sámi[4] fished for cod, halibut, salmon, herring, rockfish, and other species in summer in the great outer fjords and sometimes beyond. They hunted for seals and birds, and scavenged for stranded whales. In winter, they likely moved to the inner fjords and fished there, or in ice-covered lakes. They also hunted and trapped. They fished the great seasonal migrations of cod in the Barents Sea. Archaeologists have found old settlements near cod spawning sites, which indicates that the Sámi were there fishing for the codfish.

Sámi fishermen left other records of their existence in ancient times. They etched pictures of their lives into rock along the Arctic shoreline. In Finnmark, more than six thousand known petroglyphs of reindeer, fish, and humans with boats and fishing lines were chiseled into the stone between 4000 and 500 BC. One can see some of the "rock art" in their original locations at the Alta Museum near Hammerfest. The etchings displayed there have been designated as a UNESCO World Heritage site—a treasure of human history.

The first-known reference to the Sámi was recorded by the Roman historian Tacitus in the first century AD. He called them the Fenni, describing them as savages without weapons or horses, people who slept on the ground. Later, in AD 150, Ptolemy wrote of the Phinnoi,

Figure 16. Prehistoric rock art at the Alta Museum, Alta, Norway. The drawings depict a boat, halibut, and caribou. Used with permission from World Heritage Rock Art Centre–Alta Museum.

but these were secondhand accounts. Óttar was likely the first person to encounter and then write about the Sámi, in AD 890. Óttar was a rich Norwegian landowner and bailiff during the reign of the Viking king Harald Hårfagre. By this time there was already some domestication of reindeer. He noted that the Sámi in the north were seminomadic, fishing in summer and hunting or herding in winter. The Sámi raised reindeer for the Vikings, and they paid taxes to the Vikings in walrus ivory, seal and walrus skins, and bird feathers.

The early sea Sámi built skin canoe-like vessels to travel and fish. Scandinavian folklore is rich with tales about "Finn folk," as well as Eskimo-like people who lived in the Lofoten Islands and Vesterålen. They were also known as Skridfinnar. It's said that these coastal fishermen ventured out as far as the coast of Ireland, whether it was to explore, or to escape an enemy, or because they got caught at sea in a powerful storm. In Ireland, mysterious people with black hair, curled up inside their skin-covered and waterlogged kayaks, were mistaken as sea creatures, half man and half seal. They became known in legends as the seal folk, or selkies. Some believe that they were wandering sea Sámi.[5]

Figure 17. Portrait of sea Sámi fisherman Ivar Samuelsen in Finnmark, 1884. Photographer: Roland Bonaparte. Used with permission from Norsk Folkemuseum.

At some point, the Sámi's canoes were replaced by wooden boats; it's not known exactly when. The Sámi boat builders sewed lapstrake planks together with root-fiber cores or twisted reindeer sinews. Fishermen replaced bone and carved stone fishing hooks with iron hooks in the 1500s. They fashioned the hooks by bending and sharpening heavy gauge wire and fished using traditional hand tackle. Shortly after the introduction of iron, multiple hooks were attached to longlines. Then, in the nineteenth century, fishing nets were introduced.

Prehistoric Sámi societies developed rules and passed down traditions about fishing. The neighboring communities had reciprocal agreements about fishing in one another's territory, and "water space" was regulated. Commonly, fishing grounds were used by families and the fishing right was passed on, but with the consent of the community. The usage areas could be altered. For salmon-fishing areas, fishing rights were drawn by lot and then annually rotated among the families until the original position was restored, at which point there was a new drawing.

Historically the fjords and sea were viewed as common-property fisheries by locals. There were some rules about where and when to fish and allowable gear types. Outsiders were not invited into "owned" areas, but were accepted if they abided by the local rules.

Changing Landscape and Persecution

Torulf explained to me that there are basically three different Sámi groupings (as per prevalent and traditional ways of living): the mountain (reindeer herders), the sea (fisher/farmers), and the river (hunters) people. He believes that the sea Sámi were the indigenous people who first populated the coast when the continental ice cap covered the inland areas of northern Norway. They lived a varied way of life, fishing, farming, and hunting.[6] As the ice melted, the Sámi gradually spread inland and their culture diversified. The mountain Sámi became nomadic hunters who followed and hunted the reindeer along the migratory pathways of the beast. Extensive reindeer herding didn't develop until AD 1300 to 1500. It's thought that the mountain Sámi semidomesticated the reindeer because of the declining reindeer numbers and a desire to have a commodity to trade beyond subsistence requirements.

Over the advancing centuries, waves of settlers arriving from the south and east, driven by wars and hunger for resources, made the Sámi adapt to a changing environment and society.

The Vikings made expeditions into Sámi lands from AD 800 to 1100 to pillage, trade, and collect taxes. They may have driven the Sámi to leave the coast and settle farther into the fjords.

A so-called Little Ice Age, from AD 1300 to 1850, caused crop failures in Norway. Consequently, fishing became more important as a resource to Norse people and they increasingly populated the coast.

More settlers began arriving from the south, and the Sámi were squeezed north toward the shoreline.

In the Middle Ages, fish harvests were an important source of tax revenues for the monarchy and for local magistrates. The Black Death arrived to Norway in 1349, causing high mortality among the people. There was a drop in tax revenues as a result, so the local authorities offered the river and mountain Sámi incentives to fish, including giving them the use of some lands.

During the period from 1500 to 1900, profound changes took place in Sámi society. There were large influxes of Norse and Finns into Finnmark. Reindeer became more scarce due to hunting and loss of pasture. Inland Sámi intensified their activities of breeding and domesticating reindeer,[7] while others sought resources elsewhere, including fishing.

Settlers from the south coexisted with the Sámi, but to some extent spurned them. As late as 1900 through 1940 Norway tried to erase Sámi culture in a process called "Norwegianization." Anyone buying land had to speak Norwegian and possess a Norwegian last name. Many Sámi children were taken to missionary schools where speaking the Sámi language was forbidden.

The German occupation of Finnmark during the Second World War devastated the Sámi culture. When the Nazis left at the end of the war, they implemented a "scorched earth" policy in northern Norway. The resistance had been strong there. The Nazis burned houses, barns, and boats, and killed livestock as part of their strategy to leave the area without a military advantage. Many Sámi people were "evacuated" by the Germans; others fled for safety to caves and mountain cabins. When the war was over, the Sámi returned to rebuild their homes and lives.

Even after the war, many Norwegians looked down on the Sámi, considering them inferior and unclean people. The government's policy was to modernize northern Norway. The Sámi were encouraged to leave rural areas and migrate into cities where they could assimilate into contemporary Norwegian culture and livelihoods. Norway's social welfare state stressed equal opportunity, which tended to reinforce assimilation through economic integration. Under strong societal pressure, many Sámi rejected their indigenous culture and traditions to identify themselves as modern Norwegians. They became tax-paying wage earners instead of subsistence users of the area's natural resources.

In the 1970s an ethno-political movement began to revitalize the Sámi culture. Now it became acceptable—maybe even fashionable in some circles—to be Sámi. Public support for the Sámi identity was galvanized during protests over the controversial Alta dam project in Finnmark, which was to displace Sámi residents as well as disrupt reindeer and salmon migrations.

The Sámi Act of 1987 recognized the Sámi as an indigenous people entitled to protection and rights. Since then, Norway has been relatively progressive in terms of preserving the Sámi culture. The Sámi Parliament was established in 1989 as a representative body of the Sámi people, with special consultative powers with the Norwegian Parliament.

Most non-Scandinavian people identify the Sámi as reindeer herders (Laplanders) because of *National Geographic* magazine's feature articles about them. The articles describe a nomadic people in colorful costumes, living in tents while they follow the reindeer. However, today only about 10 percent of the Sámi live inland and move with the big herds. The total population of Sámi people across northern Europe is now about seventy thousand. Another thirty thousand descendants of Sámi immigrants live in North America.

Return of the Prodigal Son?

I had flown from Seattle to Amsterdam, then onward to Oslo, Tromsø, and finally Hammerfest. Torulf picked me up at the airport at 11 p.m. driving a middle-aged black Mercedes that was filled with tools, notebooks, and machinery. He cleared off a seat and then a space for my feet.

"Sorry to keep you up this late," I said.

Torulf waved me off. "Ha ha. It's still summer here, man. We don't go to bed."

This is the land where the summer sun never sets; it teases the horizon for a few hours before rising up to start a new day. In winter it doesn't bother to surface at all. When I arrived so late, it was still twilight in early September, but dark enough for the northern lights to show.

In the car I glanced over at Torulf, his face now and then illuminated by passing headlights. His skin was weathered and tanned, cheekbones high, coal-black hair graying. He was about my age. I noticed how similar he looked to my memory of my grandfather. (After some mental arithmetic, I was mildly startled that I was also now my grandfather's age at the point where he froze in my memory.) My family always thought our grandfather, who emigrated from Sweden, was a Laplander; my grandmother called him that when they argued.[8] The Sámi used to be called "Lapps" by the Nordic people of the south. This is considered a derogatory term among the Sámi— a racial slur. *Lapp* means both "living on the fringe" and "patched" in the Nordic languages, neither of which has a favorable connotation.[9]

After an hour-long drive from the airport, we arrived at Torulf's house overlooking Repparfjord. Sitting by the window, Torulf's eyes shifted to the fjord outside as he reflected on the history of his people. We sat talking about fisheries, politics, and the Sámi until 3 a.m. Then I tumbled into bed, exhausted.

The Demise of Sámi Fishing Rights

Conflict of the Sámi throughout the centuries with southern and coastal Norwegians extended to fishing rights. For the sea Sámi, their small traditional fisheries have been pressured by industrial fishing based in the south. Behind the oil and gas business, fishing is Norway's second-largest industry.

The cod fishery of northern Norway has been important since the Middle Ages, when dried cod was an important source of protein in Europe. The Hanseatic League (an association of northern European merchant communities to promote trade) established itself in coastal cities of Norway in the 1300s and thrived for four hundred years trading stockfish (dried cod) with the rest of Europe. Sámi and Norse fishermen worked side by side in the fishery of the northern coast.

Figure 18. Historic Nordland cod-fishing boat. This type of boat was used from about AD 950 until the early 1900s in northern Norway. Used with permission from the National Library of Norway.

By the late 1800s the nature of the fishery changed. Steam-powered vessels were introduced to the cod fishery by the British and then southern Norwegians, which enabled fishermen to trawl and purse seine with large nets. As larger ships were built, wooden hulls were replaced with iron and steel; coal-fired steam gave way to more powerful diesel engines. The Sámi often were unable to get loans from Norwegian banks, but they improved their small wooden boats—then powered by sail and oars—by utilizing small diesel engines and then outboard motors. The Sámi fishermen mostly used passive fishing techniques, like jigging and longlining, to catch cod.

More and more boats were built for the cod fishery, and eventually there was too much harvesting capacity for a limited resource. The cod population declined, and the need for management based on biological principles (such as the extent of fishing mortality, size limits, and amount of bycatch) of the species grew. By the 1960s, the large fishing boats had depleted fisheries resources in the fjords and Sámi fishermen complained bitterly, with little effect on the government or foreign trawlers. In the 1980s the local fishermen were demanding protection from the industrial boats.

Managers reformed Norwegian fisheries in the late 1980s. In 1990, a catch share program for small and medium-sized boats was adopted for cod and other species. Qualification for quota was based on past fishing harvests, while the quota quantity was based on vessel size. The Sámi fishermen complained that over those years of record used to determine qualification, catches in the fjords were low because of an invasion of seals in their fishing areas. The seals, especially in Finnmark, preyed on cod and scared the cod away from entering the fjords. Fishermen couldn't find them. As a result, the catches of cod by the small Sámi boats dropped below the threshold of consideration by the fisheries administration, and they were excluded from a full share. Instead, they were included in an "open access" group that had about 7 percent of the cod catch available to them. Later it was decided to grant a partial quota to some fishermen who were under the threshold for a full share. But a partial quota wasn't enough for most fishermen to earn a living.

The Sámi Activist

Torulf looked out the window, and his eyes narrowed to slits. Across the water from Torulf's house sits the Nussir copper mining facility, a scar in the hillside. Torulf lowered his hands to his knees, taking a wide stance, as he talked about his frustration with the mine's plan to dump toxic tailings in the water, a procedure that has been disastrous to the environment and especially to fish populations. Torulf wondered how the Norwegian government could even consider approving such a thing, but they did, unabashedly, and Torulf became an activist.

Kvalsund is near the heart of the energy and mining extraction activity in northern Norway. The Ministry of Petroleum and Energy has granted licenses to the extraction industries to drill and mine in the region. In 2011, the Norwegian Institute of Marine Research cautioned that many of the permitted areas near or in the sea should not be exploited or used as dumping grounds due to the potential environmental risk to the Arctic environment.[10] There is discord among the government agencies responsible for overseeing the exploitation of nonrenewable resources for short-term profits and those ensuring the conservation of renewables for long-term benefits.

"It is shocking that Norway allows projects of this kind," said Lars Haltbrekken, leader of the Friends of the Earth Norway, regarding a similar project where mine tailings would be dumped in a coastal fjord. After the Norwegian Environment Agency first condemned the project and then reversed course and approved it, he commented that the agency's position was "hard to understand. We thought their responsibility was the protection of our environment. The mining industry has several good alternatives available, but will now be allowed increased profits, by sending the bill for their pollution to nature, our fisheries and future generations."[11]

Even in Norway, which has a reputation for conservation, there seems to be little progress in solving the problem of industrial greed in the exploitation of natural resources. A step or two backward follows a step forward. The situation in Norway and many other countries now differs little from 1962, when Rachel Carson said we are in an "era dominated by industry, in which the right to make a dollar at whatever cost is seldom challenged."[12] Over fifty years later, things haven't changed. The Sámi hope to change Norwegian politics, and they are active in the mining issue and a number of other environmental concerns to preserve their traditional environment.

Torulf, some of his neighbors, and other sea Sámi fishermen are wrestling with the industrialists to stop the mines from polluting their fjord's water and endangering their livelihood. He calls the miners pirates because of what they take from others for their own enrichment. The company's people are dismissive and indignant about the interference of the local yokels in their pursuit of riches, and instead talk about jobs created in the mines.

Torulf also labels as pirates the big fish-trawling companies who oppose reinstating fishing rights to the sea Sámi. In particular, he mentioned one Kjell Inge Røkke. This name caught my attention because I'd written before about Røkke's business practices in Alaska.[13] Røkke, a Norwegian who had immigrated to the United States in the 1980s, founded the mega fishing company American Seafoods. Røkke never naturalized as an American citizen. In 1998, Senator Ted Stevens of Alaska inserted wording in the American Fisheries Act that prohibited foreign ownership of American fishing companies, a broadside volley directed right at Røkke. He was forced to sell. After he left Seattle and returned to Norway, Røkke founded, among other companies, Aker Seafoods ASA, which controls Norway Seafoods. Over six hundred fishing boats deliver their catches to Norway Seafoods. Another company under his umbrella, Havfisk, is reported to control 11 percent of the cod quota in Norway.[14] After promising to process the cod in Finnmark, providing jobs, it's reported that the company has been freezing the cod and shipping it to China for processing instead.

Protestors in the city of Hammerfest in Finnmark have characterized Røkke as the pirate Jack Sparrow from the movie *Pirates of the Caribbean*, who crowed the pirate toast "Take what you can; give nothing back." The fishing company counters that their corporate maneuvers save money and increase profits, and in the long run save jobs. Røkke is an astute and successful businessman, one of Norway's wealthiest citizens.[15]

As part of their struggle with Norse businessmen like Røkke, Sámi leaders have pushed for special fishing rights for sea Sámi fishermen. The Finnmark Act of 2005 legitimized Sámi land rights, but excluded fishing rights. Sámi fishermen have filed reports and staged protests, with limited effectiveness on changing government policy. The restoration of historic fishing privileges has been hampered by a lack of knowledge about the traditional fishing practices of the Sámi; it is the view of some Norwegians that only the customary practices characteristic of indigenous people should be protected—that is, the fisheries should be using bone hooks and kelp rope. Dr. Einar Eythórsson, a senior researcher at the Norwegian Institute for Cultural Heritage Research, wrote that the coastal Sámi were considered a "pariah caste" by leaders of the powerful Fishermen's Association.[16]

The Norwegian Fishermen's Association is a union that formed in the 1920s. It has great political power in Norway, especially in government actions regarding fisheries, and has opposed the concept that a group such as the Sámi should have special fishing rights. In fact, many Sámi fishermen were compelled to join the association in order to get the fishing rights available to others. In doing so, they essentially disappeared as a socially distinct group.

In 2005, the Norwegian government agreed to a consultative arrangement with the Sámi Parliament, by which they should reach common ground, including the right to fish. The Parliament created the Finnmark Coastal Fisheries Commission in 2006 to investigate Sámi fishing rights. The commission proposed that all inhabitants of Finnmark, regardless of ethnicity, had the right to fish in fjords and coastal waters, as opposed to only shareholders being allowed to fish. They recognized the collective and individual use of fishing grounds and also the right to harvest in common fisheries, but within the framework of a Finnmark-based governance of the fisheries. The commission's opinion was based on historical use, indigenous rights, and even some local court decisions going back to the 1600s, granting locals exclusive use of fish in their own fishing grounds. However, in 2011 Norway's minister of fisheries dismissed most of the commission's conclusions, ruling that the people of Finnmark had no special rights to fish in salt water. The matter is still vigorously debated and contentious.

Salmon is another historic resource of the Sámi. Fisheries managers divide the salmon harvest between the river catch (except in some large rivers where local rights holders are prioritized), which is a recreational fishery bringing in tourism capital, and the sea fishery, which is commercial and often Sámi. The proportion of salmon allocated to the two fisheries has shifted in recent years, with the amount of sea-caught salmon markedly declining. A further hardship for the fishermen is a steep decline in salmon prices, driven by the abundance of farm-raised salmon in the market. Ocean ranchers who grow the salmon in net pens are increasingly building their salmon farms in northern Norway, further threatening the future of wild runs through the transmission of diseases and parasites from the extreme concentrations of salmon in the net pens (see chap. 2). The monetary value of farm-raised salmon now exceeds that of wild-caught fish of all species, and the ocean ranchers have become a powerful political force in Norway.

A New Player in a Complex Ecosystem

Torulf introduced me to his friend Wenche during a two-hour drive to Laksfjord (Salmon Fjord). She is a Sámi fisherwoman. There's a long tradition of women fishing among the sea Sámi. Often when the men left for the run of cod in the Lofoten Islands, the women stayed home and fished to feed the family. They'd also operate the beach seines when the cod-like saithe ran, and they fished for halibut in the fjords.

Wenche's father and grandfather were fishermen. Her son, Svein, and her grandson, Matz, are fishermen. Wenche started fishing with her father when she was six years old. After she grew up, Wenche left for the south and a regular job in the city, but eventually she came back to help her son with his fishing company. Then she bought her own boat. She's strong and deliberate—like a rowan tree with roots set into the Sámi landscape. Wenche and her family fish in Laksfjord, alongside Svein's girlfriend, Alma, who emigrated from Lithuania. They each own their own small boat and fish alone. Wenche says that fishing gives her a sense of freedom. Fishing and the sea are in her blood.

After the Norwegian fisheries reform cut their share of the cod harvest in the 1990s, the sea Sámi had to find another way to make a living. The Sámi people have learned to be flexible, prospect for alternatives, and imagine possibilities in order to survive in a harsh, changing ecosystem and a shifting regulatory environment. These are tough people who've figured out a way to exist on the edge of the habitable earth by adapting.

After the fishery reforms, some sea Sámi gave up fishing for other means of living. But in the fjords of Finnmark, local fishermen found a surprising new resource—red king crab (called Russian king crab locally and Alaska king crab in the United States)—and developed a fishery for it. Wenche's family group earns the majority of their income from fishing for king crab. This species didn't exist in Norway until a few decades ago; now the red king crab in Finnmark supports a sizable fishery.

In the early 1960s a bright-eyed scientist in the former Soviet Union thought it would be a great idea to have red king crab in the Barents Sea. During the next decade, Soviet scientists shipped by rail about three thousand adults, ten thousand juveniles, and 1.5 million larvae of king crab from the northwest Pacific Ocean to the Barents Sea and released them in several inlets near Murmansk. Without natural predators and competitors, the crabs proliferated. The red king crabs have spread across the coastal areas of the Barents Sea. Recently they've been observed as far west as the waters around Iceland.

A commercial fishery for king crab in Norway started in 2002. Now about two-thirds of Wenche's income derives from harvesting crab, a resource that wasn't there for her ancestors. The crab harvest has a landed value of $14 million to $18 million to fishermen in Finnmark. Wenche's son, Svein, figured out a way to directly market his crab harvest as live shellfish to buyers in Korea, obtaining top value for his catch. Korean diners pick and choose their personal crab meal from live aquaria in fancy restaurants. Svein runs a cooperative to help other fishermen do the same.

Not everyone loves the idea of king crab existing in the Barents Sea. In recent years the United Nations partnered with several environmental groups to influence the Norwegian government to corral the invasive spread of king crab, believing that they may adversely affect the Barents Sea ecosystem. Some fishermen talk about devastating impacts of the invader crabs on the natural ecosystem, and "desertification" of the sea bottom by the voracious predators.

Yet, the crabs have also become important to the livelihood of the local fishermen, and those deriving a high level of income from fishing the crabs don't want to see them eradicated. As a compromise, in 2008 the Norwegian government established a line in the sea at 26 degrees east longitude. To the east of the line, the crabs would be managed on a quota basis with size restrictions in order to conserve them. To the west of the line, catches of the crabs are unlimited, in order to overfish them and contain their onslaught. The strategy hasn't stopped the spread of king crabs, but it may limit their abundance and thereby contain their effect on the natural ecosystems.

The Sámi Tradition Continues

Looking from the mouth of the fjord toward the sea, the sky on the horizon is gray, as if its color had drained into the ocean. The big flat ceiling stretches out as far as the eye can see and then drops down the sides of a wall to meet the water. Ancient fishermen wondered what was on the other side of that wall, the place where the winds came from and where rivers of cod swam in the waters beneath. The Sámi god of storms, named Biekagalles, lived there in the dark underworld of the north. For five thousand years the northern fishermen have survived fog and storms. Perhaps in the trough of a storm swell they caught a glimpse of the underworld. When they returned to the edge of the sea, they carved their stories on the rocks.

The history of the Sámi is a complicated braid: the background of a rugged people's everyday life, tangled with plague, invasion, famine, war, and opportunity. They have been distressed from colonization and the dispossession of their territories and resources. Even though Norway is generally recognized as a socially advanced country, the traditional rights of the Sámi continue to be denied. Since Norway is a strong supporter of the rights of indigenous people within the United Nations framework, the strategy for Torulf's group Bivdi is to use the international laws of indigenous peoples' rights that Norway has ratified in the UN to push their social agenda forward. Norway was among the first nations to approve of the United Nations Declaration on the Rights of Indigenous Peoples of 2007 and the International Labour Organization's Convention No. 169 of 1989 regarding rights of native peoples.

Sámi fishermen are few, and they are less organized than coastal fishermen with larger boats who oppose them. The Norwegian government has a history of cooperation with the Norwegian Fishermen's Association, one of the main opponents of indigenous fishing rights. The Sámi don't have treaties with the Norwegian government, but point to the above international declarations signed by the Norwegian government supporting the rights of indigenous people. The Sámi have a long fight ahead involving the legal interpretation of those rights. On the popular front, although Norwegians generally believe in equal rights for all people, the separation of a class of citizens with special rights goes against egalitarianism and the weave of the social fabric in modern Norway.

The struggle for fishing rights in Norway based on ethnic and geographic groupings has been tumultuous and complex. Torulf, the Bivdi organization, and the Sámi Parliament have had many defeats, but also some victories: They successfully pushed for an increase in the quota of common-pool cod. They have been mostly successful in keeping big trawlers out of the fjords. Boats greater than fifteen meters in length are banned from the fjords, with the exemption of purse seiners and certain fisheries.

Even with recent gains, many traditional Sámi values have been lost. Over the past millennium, the cod fishery has turned from a subsistence fishery to a trade commodity. In the last century the native language and traditional religious beliefs were mostly lost among the coastal Sámi during the process of Norwegianization. Many of the sea Sámi rejected their heritage and identified with the majority Norwegian population in order to blend in and avoid discrimination. The Sámi have interbred with Norse people for so long that they are physically indistinguishable from other Norwegians, although some Sámi say that they can "recognize their own." Perhaps the quantum of Sámi genes isn't important any longer; maybe it's living in the Arctic landscape and the choice of a way of life that now identifies the Sámi.

During my visit, Torulf mentioned that he sometimes questions who he is: a fisherman (although he hasn't been fishing in recent years)? A lobbyist? A sea Sámi? A Norwegian? How do our actions define us, as opposed to our words and our heritage? Not long after I left Finnmark, Torulf wrote to say that he had just bought himself a fishing boat. A small boat for the fjord near where his father fished. He is looking forward to hauling nets and pots in the coming winter. Once again, Torulf will read the weather and seascape, and then adapt to a changing environment—just as the Sámi fishermen have done for a thousand years before him.

Changing Fisheries

Dried cod fed the Vikings on their long voyages. For centuries the cod fed villagers of Norway through the long northern winters. Author Johan Bojer wrote a story in 1923 about traditional cod fishing by Norwegians, the way it happened for many generations. Bojer's story punctuated the demise of the old longboats in the fishery at the end of the nineteenth century, when new iron-hulled and steam-powered trawlers arrived and began to outcompete traditional Norwegian and Sámi fishermen for fish. Bojer's stirring and romantic tales linked his view of a changing world with a distant past.

When the weather turned bad and the fields were barren, Bojer wrote, Kristàver Myran took to the icy seas far to the north to feed his family. Life in Norway was rugged as the twentieth century loomed ahead. Kristàver scraped a meager living from the soil in summer as a tenant farmer of the rocky coastal fjord region, and in winter he fished for codfish. He borrowed money to buy and supply his boat. Like his father, Kristàver fished in the Lofoten Islands, "a land in the Arctic Ocean that all the boys along the coast dreamed of visiting someday, a land where exploits were performed, fortunes were made, and where fishermen sailed in a race with Death."[17] If the catch was good he could pay his bills, but if fishing was bad his family went hungry and depended on the village's social network for survival.

Kristàver harvested cod with a hook and line in a Viking-style longboat with his son Lars and six other crewmen from his village. They sailed to their destination when the wind was favorable, and if not, they struggled against the waves with oars. At the end of the day, Kristàver steered his boat to the shoreside processing plant, while the crew donned wet woolen gloves to gut the frozen catch. Their hands throbbed with the cold and the cuts in their calloused skin stung from the salt. It was hard and dangerous work.

One day a gale blew up unexpectedly, and Kristàver watched as his brother's boat capsized. There was nothing he could do in the storm to help. His brother waved goodbye, clinging to the ropes of the overturned boat, and Kristàver could only return the gesture as the wind blew him away, both realizing their paths were separating forever.

The Bering Sea Dory Fishery for Cod

In 1857, Captain Matthew Turner sailed from San Francisco to Russia on the *Timandra* with a load of general cargo when his delivery was held up outside the Amur River due to heavy ice. Some of the crewmen tossed fishing lines into the Sea of Okhotsk and hauled in large numbers of cod. Turner returned in 1864 with salt in the hold to preserve the fish caught, and delivered one hundred tons back to San Francisco, where it sold quickly. Thereby the Pacific cod fishery began, based out of San Francisco.

By 1870 there were twenty-two schooners and brigs delivering cod to the Bay Area. The fishery later shifted to the Gulf of Alaska, and the schooners off-loaded most of the harvest in Seattle, which had eight companies processing salt cod. A dozen years later, the *Tropic Bird* ventured into the Bering Sea, followed by the *Isabel*. Other schooners, like the *City of Papeete* and the *Lizzie Colby*, joined them to salt fish during summer in the icy waters. These mother ships sent single fishermen adrift in fourteen-foot oar- and sail-powered dories to catch cod. In winter, the dorymen fished from shore-based processing plants along the Alaska Peninsula and in the offshore islands.[18]

Typically, the dories would launch at 4:30 a.m. with a sail (called a *leg of mutton*), oars, a water beaker, fishing lines, a knife for bleeding the fish and cutting bait, a small keg buoy, and a windlass for hauling anchor. They would set out in a direction where they could use the wind and tides to return to their mother ship. The dorymen rowed out from the schooner up to four miles to catch the codfish, casting out a line over each side, with two baited hooks per line. The fishermen returned to the schooner at noon to off-load their catch and eat lunch. Then they would set out again. From 5 to 6:30 p.m. the dories returned to the schooner once again and were hauled aboard for the night.

Sometimes the dories were loaded down so heavily with cod that the sea almost breached the gunnels. A big cod in those days was two to three feet long and weighed sixty pounds. It wasn't unusual to catch 300 to 350 of them in a four-hour shift. The weather was usually foul. If a fisherman refused to take his dory out when the captain deemed the seas suitable, he was fined a hundred fish. If a storm blew in before the dories could get back to the mother ship, they might disappear in the mist, sometimes never to be seen again. In such cases the captain would tie a line to an empty dory and let it drift behind the schooner one half mile or more in hopes the lost boat would run into it.

In 1909, the schooner *Harriet G* launched twenty dories in a light breeze. The wind shifted direction as it began to grow in strength. Blown away from the mother ship, nine dories were lost in a single day. One dory was later recovered by another schooner, and the other eight washed ashore. Only two of the beached fishermen survived; they made a long overland hike through grizzly bear–infested tundra to reach a lighthouse from which they were later rescued.

Norwegian immigrants manned most of the dories. Often they were novice fishermen. Some of the best fishermen working for the shore-side processors were Native Americans. The top fishermen would harvest salmon and halibut, which were more lucrative than cod. It was a hard life in a harsh environment. Fishing trips lasted three to five months, and when the fishermen arrived back on the mainland, the money they earned at sea could disappear quickly on a few earthly pleasures.

In 1927 motors were installed on the dories, ending the era of wind and muscle power. The last schooner, the *C. A. Thayer*, participated in the Bering Sea dory cod fishery until 1950. The *Thayer* was restored to the ship's original configuration as a log carrier and now rests at anchor at the San Francisco Maritime Museum as the last of its kind.[19]

RETURN TO ARTISANAL

To stand at the edge of the sea, to sense the ebb and flow of the tides, to feel the breath of a mist moving over a great salt marsh, to watch the flight of shore birds that have swept up and down the surf lines of the continents for untold thousands of years, to see the running of the old eels and the young shad to the sea, is to have knowledge of things that are as nearly eternal as any earthly life can be.
Rachel Carson, *The Edge of the Sea*

The use of traditional methods to harvest fish, the development of markets for underutilized species, better processing of bycatch, organization of community-based fisheries, direct marketing, and alternative forms of managing fisheries are elements of modern artisanal fisheries.

Every artisanal fishery faces unique problems and challenges to its survival. This section presents several examples. A Salish Sea reefnet fishery uses traditional capture technology but struggles from a shortage of fish (chap. 6). A Nova Scotian weir fishery, also using traditional capture methods, is threatened by hydroelectric power companies (chap. 7). The Puget Sound gillnet fishery for chum salmon is a low-impact fishery of an underutilized species and uses direct marketing from the fisherman to the consumer (chap. 8). The fishery competes against larger and more powerful purse seiners and struggles with diminishing salmon runs caused by pollution, habitat destruction, and hydroelectric dams.

6: A Clean and Green Fishery

LEGOE BAY REEFNETS

Skipper Ian Kirouac pressed his cell phone to one ear and a radio to the other while he barked orders to his crew from atop his perch. Meanwhile, he kept a sharp eye out for salmon approaching his trap, multitasking a mix of traditional and modern technologies. Suddenly Ian's gaze focused hard on the water. He spotted fish.

"Take 'em," he yelled, and all hell broke loose.

Ian yanked on a rope, releasing an armada of whirring gears; the tail end of the net lifted out of the water, and two fishermen pulled frantically hand over hand on the mesh to isolate a school of sockeye salmon in a corner pocket. They lifted up on the netting above the fish and the sockeye splashed into a live tank belowdecks. The crew members were left gasping for breath.

This is the reefnet fishery in Lummi Island's Legoe Bay. The Washington Department of Fish and Wildlife calls it the best selective fishery around, and the Monterey Bay Aquarium's Seafood Watch gives it a "Best Choice" rating. The reefnet fishery in the Salish Sea employs energy-efficient mechanics, clean harvesting techniques, a lean strategy for their supply chain, and state-of-art management to operate one of the best-practice fisheries on the planet.

While I observed the fishery for a day, about six hundred sockeye and pink salmon were caught. It was only the beginning of the season—later on, a single haul might net that many. One starry flounder and one king salmon were intercepted as bycatch; neither was damaged and both were released back into Puget Sound without harm. Tests have shown that the mortality of released fish is less than one-half of 1 percent.

Figure 19. A man on a perch watching for salmon to enter the reefnet set, Legoe Bay, Washington. Photographer: K. Bailey.

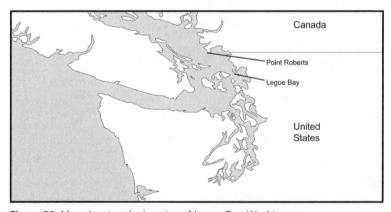

Figure 20. Map showing the location of Legoe Bay, Washington.

Energy-Efficient, Well-Managed Fishery

A "gear set" is a pair of stationary platforms anchored along traditional migratory routes of the salmon on their way back to the Fraser River to spawn. No fossil fuel is burned to chase these fish; they come to the fishermen. The winches are powered by solar-charged batteries, and the rest of the energy is supplied by muscle power.

Based on a preseason test fishery and population models that forecast the number of salmon returning to rivers, scientists set an allowable harvest level that will leave enough fish behind to conserve the reproductive capacity of the salmon populations. Simulation models are then used to determine gear operating times in the water at different locations to approximate the allowable harvest. Fishermen take only what's above and beyond that needed by the population to sustain itself at a healthy level.

As a further control on the fishery, before the harvesting is even allowed to begin, genetic samples from test fishing are analyzed to ensure the flow of migrating salmon contains only a minimal number of fish from stocks (or reproductive groups) that might be depleted. There are nineteen recognized Fraser River spawning stocks; some of these are protected. In 2015, for example, the Chilko lake-type and late-run Shuswap stocks were expected to dominate catches. The late-run Stuart and Cultus stocks were protected.

Unlike fisheries whose strategy is to catch as many fish as they possibly can, and then try to get rid of them like a surplus of Lincoln Navigators on a car dealer's lot, these reefnet fishermen utilize a lean supply-chain approach and catch only enough fish to fill orders on hand.

Modification of a Traditional Fishery

Centuries ago, probably between AD 1300 and 1500, the indigenous people of the Pacific Northwest perfected the technology of reefnets. The people of the Northern Straits Salish, from southern Vancouver Island and British Columbia to the San Juan Islands and includ-

Figure 21. Fishermen retrieving salmon from a reefnet, Legoe Bay, Washington. Photographer: K. Bailey.

ing the Lummi and Samish tribes, were particularly adept at reefnet fishing. The Lummi contend that their ancestors developed the reefnet fishing technique, and the technology was transferred to other tribes through marriage linkages. The ability to catch large amounts of salmon, coupled with a mastery of preserving and storing the fish, enabled their cultures to thrive.

The Lummi's historic fishing spots were found at the locations now called Birch Point, Cannery Point on Point Roberts, and Sandy Point in the northwest corner of Washington. Reefnets were also fished in the waters off Lummi Island, Orcas Island, and San Juan Island. The placement of nets took advantage of strong tidal flows and the natives' knowledge of the ocean currents and migratory routes of salmon. The Lummi reefnet sites were territorial, but other fishermen were granted permission to fish at an occupied site in exchange for a portion of the catch. The Point Roberts reefnet site was shared with the Semiahmoo, Saanich, and Cowichan tribes.

The ancient reefnet, made of willow saplings, the inner bark of cedar, and sometimes nettles, was laid on or near the bottom and sloped toward the surface at the downstream end of a flood tide. The net was attached to anchors at the upstream end of the surging tide and to canoes crewed by five to ten men at the other end. Anchors kept the canoes apart at the downstream end to spread the net. A channel into the net was created with rope leads. Seaweed fronds and dried grasses were woven into the side and bottom ropes to lure the salmon into the waiting net. Like sirens, the fronds of kelp braided in the side netting beckoned and channeled the fish. The false bottom caused the salmon to rise toward the surface. There are accounts of the men clearing away rocks on the bottom to make a channel and also of clearing a path through the kelp to guide the fish to their nets. When spotters observed salmon swimming into the net, the canoes were released from their downstream anchors and swung together in the current. Meanwhile, the downstream end of the net was pulled in, followed by the upstream side to trap the fish. The lines would grow taut as the salmon in the belly of the net convulsed and twitched, struggling to get free.

These days, the operation is similar to the ancient technology but with a few modifications. The nets are now made of lightweight nylon. Plastic streamers have replaced seaweed and grasses. A tower is constructed over the net to observe the fish. Solar-powered winches supplement the fishermen's muscle power.

When I visited, the collection of gears, electrical switches, and pulleys on the deck of Ian's offshore platform reminded me of my machinist grandfather's garage in the 1950s. A few electric motors, scavenged from hatches of World War II bombers, are still in use in the fishery. It's a complicated setup to first pull the head of the net up to stop the incoming fish, then pull the bunt, or rear of the net to trap them, and finally lift the bridge, or middle, to concentrate them in a pocket.

Figure 22. Drawing of Coast Salish reefnet gear. The net is anchored with stones at one end and held open by canoes at the other. The sides of the net are braided with kelp fronds. When fish enter the trap, the canoes draw together, closing on the fish, and the net is hauled in. Artist: Mattias Bailey.

"Give 'Er Hell"

Legoe Bay, just west of Bellingham, is long and thin. Its beaches are sandy, strewn with driftwood. The white cone of Mount Baker stands high in the distance. There's little sound on the water among these platforms—just water slapping wood, ropes creaking, and the chatter of fishermen on the adjacent gear sets.

I watched as an old dog paced the deck on one of the gear sets, keeping an eye on things. When fish arrived, the quiet was broken by war whoops and the clatter of winches being deployed. Ian's partner, Sierra, sat in a shed on the inshore platform as she monitored the displays of four up-looking video cameras mounted on the bottom of the net. Two other crew members, Jake and Josh, did most of the net hauling from the offshore platform. The net was submerged between the two platforms.

Ian hollered, "Sierra, give me some breast."

I looked over at Josh, my eyebrow raised as a question, while my other eye strained in its socket to watch Sierra—strictly out of professional curiosity. Josh is a tall, hardworking Texan whose easy drawl reminded me of the movie star Matthew McConaughey. He smiled and told me that if he were in the shed, Ian would use the same words, meaning to let out some line and square the net.

From the spotter's position, the incoming salmon can first be seen as a fleeting shadow or even a ripple on the surface from about 150 feet out. The advancing fish take shape as ghostly silver forms gliding through the dark water. As the school closes on the net, the spotter's adrenaline surges. Sitting up in his perch, he's like a great blue heron on the hunt. Sierra confirms from the cameras that the fish are trapped. "Give 'er hell" is the traditional call to haul the fish. When the nets are pulled, the tower sways in a manner most greenhorns, like myself, find unnerving.

After the harvested salmon are netted, they are kept in a live tank, then bled in the water and iced in a slurry until delivered ashore, where processors fillet and debone them by hand. The product is the highest quality.

I mentioned to Ian that a Lummi fisherman told me about how his ancestors would leave a hole at the working end of the net to guarantee that some fish would escape the nets and return to spawn in the river. (The fisherman said it was called the "vagina." He stared at me intently as he said it. Was he pulling my leg? Is there a thing about these nets and female body parts? I could imagine writing about the planted symbolism and then looking like a fool. Later I found an outside reference that said the Lummi considered their nets as females; the hole was called the vulva.[1] So—pretty close, anatomically speaking.) Ian looked at me sharply when I told him about the hole. "I'm not sure I buy that. If there was even a little hole in this net, a fish would find it and the rest of them would follow behind. The net would be emptied in seconds," he said.

Sometimes there are differences in the traditional knowledge and the Western viewpoint that are hard to reconcile. Maybe the hole in the Lummi net was a relief valve during an era when the fish were more abundant. I imagine the nets could become too heavily laden with fish for mere muscle power to pull them, and a hole with a closing mechanism would be a way to let excess fish spill out.

Reestablishing a Responsible Fishery

Although Straits Salish tribes, including the Lummi, developed the reefnetting technique, currently no Lummi fishermen operate commercial reefnets in Legoe Bay.[2] Instead, they fish for salmon with purse seines or gillnets. Fishing is still a strong tradition in this tribe, particularly among the older people. As one Lummi fisherman said, "Fishing is who we are."[3]

In the 1890s the Indian reefnets disappeared through a combination of competition with non-Indian salmon traps and adverse court rulings on land use. The first traps set by non-Indians creeped into the Point Roberts fishery in 1892. By 1894 there was a complete line of traps across the fishing grounds, catching the salmon before they arrived at the reefnets, thereby rendering them useless.

The land around the main traditional fishing site at Cape Roberts was "homesteaded" by a white settler, who blocked the Indians' access. Some reefnetting continued off Lummi Island until the last local cannery closed in 1924. For a while, that was the end of reefnetting except for subsistence fishing.

When salmon traps were outlawed for all fishermen in 1935, hundreds of non-Indians set up reefnets to replace the traps. They hired Lummi fishermen to show them how to use the gear. In 1939 a new cannery opened to buy reefnet-caught fish. A few Indians also put in their own reefnets, but they had been crowded off the best locations to intercept fish and didn't succeed. Because of the fixed migration routes of sockeye salmon and their interplay with tides, the location of the nets is critical.

At the high point of the non-Indian fishery, seventy reefnets worked in Legoe Bay alone, but now they have been whittled down to a reasonable number to match the harvest the fishery can sustain. Eight gear sets remain in Legoe Bay, and three more fish in the outer San Juan Islands. Four of the Legoe Bay licenses belong to the Lummi Island Wild Cooperative. The gear sets of the co-op share their catch quota; crew members shift from platform to platform, depending on where they are needed.

In 2007 Lummi Island Wild won the Governor's Award for sustainable practices. In spite of the reefnet fisheries' green approach, their survival isn't a sure thing. Populations of the most valuable species are sliding downward in the northern Salish Sea. There is no king salmon fishery now, due to the low numbers, and in 2016 the Fraser River sockeye fishery was closed due to a poor run. In the marketplace, the craft-caught salmon of the reefnets competes against high-volume harvests of purse seiners that sell for a cheaper price.

Lummi Tradition and Activism

The historic fishing areas are still important to Lummi fishermen, although they mostly use modern gear like purse seines to fish there now. The Lummi tribal fishery is the largest native commercial fishery in the United States, in spite of declining participation. In the 1980s the Lummi fleet numbered up to 700 fishing boats, but now it's down to about 450 vessels.

The decreasing number of salmon in their customary fishing spots alarms the Lummi fishermen, and they have turned to activism to protect their heritage. The Lummi Tribe led the opposition against building the Gateway Pacific coal terminal at Cherry Point in the heart of their historic fishing grounds. Local non-Indian politicians and the energy-extraction industry promoted the benefits of employment at the terminal. But the opposition held firm in the belief that their fisheries would suffer long-term consequences, resulting in a loss of traditional jobs. A study by the Port of Bellingham in 2013 showed there were 2,816 jobs through the port related to fishing, while the coal terminal would create 1,251 jobs.[4]

In May 2016, the Army Corps of Engineers sided with the Lummi and denied the coal terminal a permit to build as a violation of Indian fishing rights. The Lummi's next goal is to restore the Cherry Point herring population, whose decline trailed closely behind the construction of a smelter in 1966 and an oil refinery in 1971 at the Cherry Point site. The herring themselves were a valuable fishery; but even more critical, since king salmon depend on herring as prey, the herring's demise also decimated the area's king salmon fishery. Between 1973 and 2012 the Cherry Point herring stock plummeted 92 percent. Estimates of Chinook abundance in the Puget Sound are as of this writing about one-tenth of their levels in the early 1900s. Chinook salmon in the Nooksack River of Bellingham Bay are now listed as critical under the Endangered Species Act. Although the diminished Chinook population has suffered many stresses, from pollution to habitat destruction and dams, the lack of herring in their diet is probably a major factor as well. A large disturbance at one end of the ecosystem rattles the other end. Officials of the chemical industries blame "the environment."

Tribal elders from the Lummi Nation haven't given up on the reefnets. After more than a century of absence, Lummi fishermen attempted to reestablish reefnetting in traditional locations as a youth demonstration program in 2013 and 2014. They weren't successful in catching salmon. It will take time to relearn the old ways and train a new generation. In 2016, with technical assistance from Lummi Island Wild fishermen, the Lummi tribal fishermen set up modern reefnet gear near Cherry Point. It's as if history is replaying itself in reverse, from when tribal fishermen once helped the nontribal fishermen.

Chef's Seal of Approval

Lummi Island reefnetted salmon is a community-supported fishery, meaning that consumers buy salmon directly from the fishermen through local buyers' clubs. The product can also be found in finer markets and restaurants of the Pacific Northwest that support sustainable fisheries.

A number of award-winning chefs at celebrated restaurants put Lummi Island reefnet-caught salmon on their menus. A *Food & Wine* magazine Best New Chef of 2012 and a James Beard Foundation's 2014 Rising Star Chef of the Year, Blaine Wetzel (of the Willows Inn on Lummi Island), has said, "This is a fishery that I can stand behind. I love the flavor of the fish; it's the highest quality."[5] I talked with John Sundstrom, the chef/owner of Lark restaurant in Seattle who was named Best Chef Northwest by the James Beard Foundation in 2007. He said, "It is an excellent fish. I can ask them when the fish came out of the water and they can actually tell me. No other salmon is being sold this way. It's a fantastic way to catch salmon, and the treatment of the product is superb."

The modern reefnet story illustrates an artisanal approach where fishermen harvest salmon using traditional principles to attain a clean catch. Management of the fishery is state of the art. At the same time, the harvest technique has been modernized for energy efficiency.

"Sea Gypsies" and Fish Listeners

Some unusual types of artisanal fishing cultures are dying out, including the sea gypsies, or Moken, of Myanmar, and the fish listeners of Indonesia.

The Moken are a seminomadic group of Austronesian people who live on wooden boats most of the year. They live off the coast of Myanmar in the turquoise water of the Mergui Archipelago, which comprises about eight hundred islands spread across 250 miles in the Andaman Sea. The Moken come ashore during the monsoon season, but otherwise live on their boats for eight or nine months a year, collecting fish, shellfish, and anything else they can to eat or to barter with Chinese and Malay traders for rice and gasoline.

The Moken travel in a flotilla of seven or more boats called *kabang*, forming an extended family. The kabang are made of wood and bamboo, lashed with rattan rope and thatched with dried palm leaves. The fishermen catch their prey with spears, harpoons, nets, and handlines; they dive for shellfish on the bottom or collect them from intertidal pools.

The Moken have been persecuted, harassed, and exploited—and their numbers are declining. The Myanmar government has tried to contain them in a national park as a tourist attraction. There are only about a thousand people of the tribe left, and many now live in permanent land-based villages. Western observers report, "The separation from their ancestral environments and traditions is, more often than not, catastrophic for the long-term mental and physical health of tribal peoples such as the Moken."[6] The Moken people themselves say, "The Moken are born, live, and die on their boats, and the umbilical cords of their children plunge into the sea."[7]

. . .

The fish listeners of the Setiu Lagoons in Malaysia hunt for fish by diving in the water and listening for sounds. It is a knowledge passed on from father to son in order to hear fish in the water from up to three kilometers away. As the fish listener hunts, he is accompanied by a pack of speedboats; when he locates a school of fish, he signals the boats and they drive the fish into waiting nets.

One fish listener put it this way: "They [the fish] have a voice. This sound is this fish, that sound is another. When someone is new, they can't tell one fish song from another. . . . After a while, it is as if you can see. Even though the fish is very far, you can sense it in that direction and you go there. Only when you get close, you can hear the fish clearly."[8]

Between 1971 and 2007 Malaysian fish stocks declined 92 percent due to overfishing, according to one fisheries expert.[9] With the decrease in fish stocks and the loss of traditional knowledge, the fish listeners are a dying breed.

7: Crimson Tide

THE BAY OF FUNDY WEIR FISHERY AND A CONFLICT WITH GREEN POWER

"I am a fisherman," Darren Porter said in a thick Scotian accent. "It's not only what I do, but who I am."

He is big and burly, built like an oak stump. If I was in a bar fight, I would gladly have Darren in front of me to clear the way. He operates a weir fishery in Minas Basin on the southeast side of the Bay of Fundy. The bay has the highest tides and strongest currents in the world, which leads to a problem for Darren. The power industry wants to install giant turbines in the passage to Minas Basin, maybe over a hundred of them, to harvest the wealth of Nova Scotia's tides, generating megawatts of energy and enormous profits. The turbines look like giant food processors, five stories high. The power industry also has a problem, and that is Darren.

The Bay of Fundy is at the end of the Gulf of Maine, bordered by the provinces of New Brunswick and Nova Scotia. The southern terminus of the bay is Minas Basin. When the tide flows into the outer bay, 160 billion tonnes of water come in at a speed of one to two meters per second. Where the bay narrows to squeeze through the five-and-one-half-kilometer-wide Minas Passage, fourteen billion tonnes of seawater accelerate to five meters per second. It's claimed that the volume of water transported is more than the flow of all the world's rivers combined. But it is the speed of the flow that makes the area attractive to the energy industry.

For Darren, every tide is either "Christmas—or a slap in the face," depending on how many fish it brings with it. When I visited the weir with him in June 2016, as we drove up he eyeballed the gulls and herons gathered around to feed on the fish trapped there, then lit his

Figure 23. Fisherman Darren Porter at his weir in Minas Basin, Bay of Fundy, Canada. Photographer: K. Bailey.

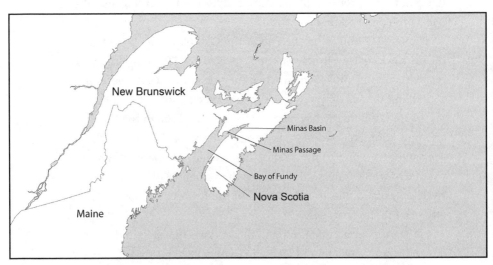

Figure 24. Map showing the location of Minas Basin, Bay of Fundy, Canada.

pipe and said in a broad voice, "We got fish in there today." I found it hard to take notes as the pickup truck bounced up and down over rocks; it was like trying to drink tea on a galloping horse.

Darren's weir is a type of fish trap that uses the tides to bring fish into it. The design is ancient, having been used by the First Nations people of Nova Scotia well before the region was colonized by Europeans. In the recent past, there used to be a weir every mile, maybe a couple hundred of them. Now there are only six.

Darren's weir has two wings that are about three hundred meters in length, constructed with one thousand poles. Placement of the walls of the weir is critical in determining which species he will catch in the trap. He wants to avoid striped bass, salmon, and sturgeon. The weir is covered by netting to guide the fish into the trap section where the two wings come together. At high tide Darren's weir is covered by ten meters of water. When the tide goes out, the weir goes to work, trapping the fish behind the netting. One wing of the weir ends in a candy cane to trick the fish from swimming around the last pole and out of the weir. Darren figures that most of his fish are caught in just twenty minutes of action as the tide recedes. There is a pond in the trap to keep fish alive, and a gated channel from the trap to a recovery/holding pond, shaped like a heart, where unwanted species are held until the next tide comes in and takes them away.

I was accompanied to the weir by Mike Dadswell. Mike is a retired professor and fisheries biologist. He still has projects and students working with him. On this day he was measuring sturgeon that were caught in the weir. Then he attached small, fluorescent, spaghetti-like tags to the fish and released them. When the tagged fish are caught again, remeasuring them yields information on their growth and movements between capture events. Mike's knowledge of the Bay of Fundy, the fishes, and turbines is encyclopedic. He told me that Minas Basin is an environmental treasure and a sensitive ecological site. Because of the warm waters of the bay, it is a cul-de-sac terminus in the range of several species, like shad, moving up the East Coast from as far away as Florida.

The biodiversity of the fish in the Bay of Fundy is relatively low; most species can't cope with the turbulence of the extreme tides. Fifty

Figure 25. Darren Porter's weir in Minas Basin, Bay of Fundy, Canada. Photographer: K. Bailey.

species of fish are caught by Darren's weir. (By contrast, 253 species of fish live in Puget Sound.) But what is living in the Bay of Fundy is abundant. It is one of the most important American lobster nurseries, and it seeds the Gulf of Maine with young lobsters.

Standing by the weir, I heard a yelp and a laugh from the table where the fish retained by the weir are sorted. Darren's daughter Erica, who works the weir with him, grinned and held up a sculpin to show me, joyous that a different species broke the monotony of their usual catches.

The Porter weir fishes from mid-March to mid-August. Darren doesn't make much money—just enough to survive. He's forbidden to keep the most valuable fish, the ones that were the staple species for fishermen's lives in the old days. Canada's Department of Fisheries and Oceans (DFO) has closed all weirs in Minas Basin for Atlantic salmon, striped bass, and sturgeon. Catches were declining and the government decided there was more money in recreational fishing, especially for salmon and bass. These days Darren sells some flounder to restaurants. In previous years, a fish distributor took his catch of fish and marketed it. But fish distributors come and go, and that guy's gone. In the distant past there were fish peddlers who would take the catch from the weirs and sell the fish door to door. They're long gone. Now most of the herring, flounder, shad, and gaspereau (aka alewife) are sold as bait to lobster fishermen. Darren's not allowed to process the fish he catches by government laws; he can only harvest and ice them. He complained about the restriction and how it affects his ability to add value to his product.

A FORCE in the Bay

Fundy Ocean Research Center for Energy (FORCE) is a nonprofit organization formed by energy developers and the Nova Scotia Province's Department of Energy. The role of FORCE is to serve as a host to technology developers, a catalyst to safe development of power. They provide the infrastructure to deliver tidal power to the grid, and the organization also oversees "independently" reviewed environmental monitoring. To date, FORCE has installed a 16-megawatt subsea power export cable at a cost of $15 million. They are already renting space on the cable to partners: Cape Sharp Tidal, Black Rock Tidal Power, DP Energy, Minas Energy, and Atlantic Resources. The FORCE board of directors has six members from the power/engineering corporate world, one biologist from Acadia University, and finally, the owner of a public relations company that represents Nova Scotia Power in such matters as rate increases to the public.

FORCE has claimed that the Bay of Fundy has the potential of producing 7,000 megawatts of power, enough to provide for three million homes. The executive director of the Offshore Energy Research Association of Nova Scotia, Stephen Dempsey, has cited studies that tidal power in the province will create twenty-three thousand jobs and generate $1.7 billion of revenue.[1]

In summer of 2016, FORCE wanted to install two "test" turbines

Figure 26. The OpenHydro-built turbine installed by Cape Sharp Tidal. The height of the turbine is five stories. The turbine was installed in Minas Passage. Used with permission from Adam London.

for Cape Sharp Tidal, a cooperative venture of Emera and OpenHydro. Emera is the owner of Nova Scotia Power; Nova Scotia Power provides 95 percent of the power to the province. Emera, an international conglomerate, has $11 billion to $12 billion in assets in Maine, the Caribbean, and the Canadian Maritime provinces. One turbine was installed in 2016; the other (as of this writing) has been delayed.

Cape Sharp Tidal's web page[2] indicates their plan is to generate 16 megawatts in 2017 (eight turbines), 50 megawatts in 2019 (twenty-five turbines), and 300 megawatts by the 2020s (150 turbines). Black Rock Tidal is planning on forty turbines. The number of turbines planned by the other three companies is unstated. But with just a hundred of them, every drop of water in the basin would pass through the blades of the turbines each month.

The companies involved in tidal-energy production in the Bay of Fundy, the nonprofits supporting them, and provincial and federal agencies regulating them are intertwined. The provincial Department of Energy's mandate is to promote the energy sector, facilitate development, and regulate it. They have the tough goal of getting Nova

Scotia off fossil fuels and onto renewable energy. The provincial Department of the Environment has the job of making sure that the environment isn't damaged by coastal development projects. DFO plays an advisory role, but regulates takings of animals listed as endangered or threatened species. FORCE plays a dual role; besides facilitating the turbines, the organization oversees environmental monitoring.

FORCE's environmental effects monitoring program, or EEMP,[3] was assessed by the federal DFO and the nonprofit Ecology Action Centre (EAC). Letters written by EAC and DFO in 2016 pointed out, among other issues, the inability of FORCE to monitor mortality of marine animals encountering the turbines, the general lack of knowledge about the interaction of marine resources with the turbines, the lack of accounting for the effects of scaling up when more turbines are added, and the fact that the technology employed in monitoring is incapable of species identification. EAC wrote, "A lack of existing technology is not a sufficient excuse to not properly monitor effects."[4]

In spite of these problems, the Department of the Environment signed off on FORCE's monitoring plan, calling it "an adaptive monitoring plan." In other words, a monitoring plan that isn't ready or complete, and that will change over time.

Margaret Miller has been the reigning minister of the department since January 2016. Her previous experience prior to election to the Legislative Assembly in 2013 was helping to run the family dairy farm and then managing a small woodlot. She was reported to say about the effect of the turbines on fish and shellfish: "Resources evolve."[5] Then in response specifically to the effect of turbines on sturgeon in the bay, she said, "Sturgeon? Do we make money off of those?"[6]

According to EAC, FORCE's monitoring plan was not released publicly because FORCE said they were "unable to share commercially sensitive information with the public."[7]

It's a conundrum. FORCE can't determine the impact of the turbines at all until they are installed and operating, and the technology to determine the effects isn't good enough to monitor the impact. The ability of new acoustic technology to detect fish and identify them in the rugged conditions of Minas Passage is unknown. There's another problem with the monitoring: even if one could determine abundances of animals before and after turbine installation, the natural variability in the Bay of Fundy—the changes in abundance of fishes due to annual, seasonal, daily, intertidal, and spatial patchiness effects—will likely overwhelm the variation caused by one or two turbines. It seems inevitable that since FORCE can't detect an impact, permits for new turbines will continue to be issued. But the turbines will have a cumulative effect, which will be magnified each time a new turbine is installed.

A Clash of Cultures

The installation of the turbines in the Bay of Fundy represents a collision of two worlds: independent fishermen against engineers and corporate opportunists.

Darren Porter said that the monitoring is being developed as a "test as you go" approach. "It puts the province so deep into the project, I fear they will proceed with it no matter what ecological disasters will be predicted to flow from it," he said.

"We are often overlooked by the excitement of a new industry wishing to share the marine environment with us, but find ourselves having to become fishermen scientists and guardians of the environment," Darren said in a newspaper quote.[8] He pointed out in another report, "We haven't been able to kill a striped bass since 1996, not commercially, so I want to know how many striped bass they are allowed to kill? How many is acceptable? How many harbor porpoises?"[9]

Mike Dadswell commented on the turbines, "Why sacrifice one renewable resource for another."[10]

Matthew Lumley is the communications director of FORCE. I met with him in their tenth-floor conference room in downtown Halifax. Before I could ask a question, he started the interview with the suggestion that perhaps he should just tell me about the project.

He began by recounting how there is evidence of tide mills in the Bay of Fundy from the early 1600s. People have tried to harness the tides for power ever since. In 1915 a physics professor named Ralph Clarkson at Acadia University built a prototype turbine he planned to set up near Cape Split at the entrance to Minas Passage. The project became the Cape Split Development Company. But the device was destroyed in a fire, ending Clarkson's dream. Then a Passamaquoddy Bay landowner named Dexter Cooper talked with his neighbor, Franklin Roosevelt, about the great potential of tidal power and it grabbed Roosevelt's imagination. Roosevelt's government funded the building of a dam to harness the power, but Congress later defunded it. Over time, the concept of tidal power has folded in and out of favor. (Just before his assassination, John F. Kennedy talked about the potential of tidal power in a speech.)

I sensed Matt was trying to establish a precedent for the use of tidal power in the bay. Something he doesn't mention is a turbine installed in 2009 whose blades were torn off. A dead whale was found nearby just after the turbine was broken. The coincidence has generated discussion of whether both the turbine's and the whale's demise was the result of a collision between them.

A turbine venture Matt briefly mentioned is the Annapolis Royal project, a 20-megawatt plant that was built by Nova Scotia Power in 1984. Matt explained that the Annapolis turbine complex is a "barrage" type, different from the freestanding turbines planned by Cape Sharp Tidal. A barrage turbine traps water behind a dam on the incoming tide, and captures power when the water is released.

The Annapolis Royal head pond is believed to have captured two whales, one of which died. Mike Dadswell calls it a "fish killer." A once-healthy striped bass population that ran up the Annapolis River has disappeared since the turbine was installed. There is controversy about whether the turbine or the dam destroyed the fish run, but it's not even an endangered stock anymore; it no longer exists. There is no controversy on the damage the turbine has inflicted on other fish populations that spawn in the Annapolis River.[11] In spite of the environmental damage, the Annapolis Royal plant is still producing electricity for Nova Scotia Power. Even though it is still considered a "test project" after thirty years, they have chosen not to remove it, nor has the Department of the Environment taken action.

Matt carefully explained how FORCE is a nonprofit organization whose six employees do not work for private companies; they are passionate about exploring the technology of tidal power. He pointed out that Nova Scotia is a particularly high user of fossil fuels and an emitter of greenhouse gases. He said, "Let's start now and look at the technology. We can't not ask this question."

To begin with, Cape Sharp Tidal plans to install two 2-megawatt test turbines. It's the speed of the tidal current that attracts the industry engineers, since power density is a function of the cross-sectional area times the speed cubed. The energy increases exponentially with the speed.

Matt said, "We have common ground with the fishermen; we all have questions. The two central questions are: Is it safe, and is it affordable?"

The new technology of "in-stream" turbines is different from that used at Annapolis Royal. The new turbines weigh one thousand tonnes and stand fifteen meters high.

"FORCE concludes that any harm by the turbines to fish would be insignificant. How much do you have to study before you put one in?" he said.

Matt told me how the new turbines are designed to run at a slow speed—six to ten revolutions per minute—and fish will avoid the turbines and the blades. After all, he points out, "Fish avoid rocks."

Hearing him say this, I think to myself how six to ten RPM seems slow, but each turbine has ten blades. A blade slices by any fixed point at a rate faster than one per second. Using a combination of vision and motion-sensing cells in their lateral lines, fish can avoid objects. But in swift-flowing muddy water, blades passing by every second could present a challenge. Shear and pressure effects of the rotating blades can also damage fish. Mike Dadswell pointed out that the current in Minas Passage flows at five meters per second, by far exceeding the average swimming speed of most fishes. At those speeds the fish don't control their interaction with the turbines, the currents do.

Lumley was so enthusiastic about the turbines and confident about the lack of danger they impose, my curiosity was aroused. I asked him about his professional background. Although I was expecting to hear about his training in environmental science, he answered, "Theater and journalism." My surprise should have been at my own naïveté; after all, he is a communications director.

I recognized a lot of Matt's wording as coming directly from an opinion piece published a few days earlier in the *Chronicle Herald*. It was authored by the general manager of FORCE, Tony Wright.[12] That's Matt's job as communications director and he does it well, coming across as persuasive and friendly, and on point.

I looked at the whiteboard in front of the room. Written on the board:

Why are we doing this?
Economic opportunity
Clean energy future

It looked like a playbook, or perhaps more aptly, a carefully crafted script. Wright's background is in naval engineering and business. I didn't see any mention of the environment of the Bay of Fundy.

An Uncertain Future

As we drove off the tideflats in his pickup truck, Darren turned to me and asked, "Whose water is it?" Then he told me a story from the First Nations Mi'kmaq people about how the tides in the Bay of Fundy started. Glooscap, the man-god creator, wanted a bath so he ordered the beaver to build a dam across the mouth of the bay. Then the whale came and Glooscap recognized that the whale needed to pass through. Life needs to pass through. So he ordered beaver to breach the dam, but the beaver was too slow, so the whale broke it with his tail. The slapping of the whale's tail formed the tides, which continue to this day.

Recent scientific studies have shown truth behind the legend. The tidal flows in the bay were initiated by the breaching of a natural blockage about six thousand years ago. It's amazing that the event has been preserved in a story for so long.

Most of the flow and marine life is directed through a narrow channel in Minas Passage. The area is frequented by whales, sharks, seals, lobsters, herring, shad, sturgeon, sea bass, and salmon. It is exactly here that the power companies want their turbines deployed.

Darren said that he always wanted to be either a fisherman or a biologist. His family farmed clams, mussels, and oysters. He started out in the fish business by clamming. Later he began fishing with gillnets for small pelagic fishes. He bought his weir six years ago.

"Fishermen have a personal connection to fish. They handle them every day. People say we are fish killers. But I can't look a fish in the eye when I take its life," Darren said.

Darren is studious and knowledgeable about fisheries science, the ecosystem, and the turbines. He said that he has gone through some of FORCE's formulations on strike probabilities and found errors in their calculations. He's pointed these out to FORCE, and has been infuriated when they keep citing the same wrong numbers in meetings, misinforming the public. He's also angry when they quote studies from Scotland saying that turbines show no evidence of damage to marine life. What they leave out is the part that says that the technology to detect damage isn't available. "Lack of evidence of impact is not the same as evidence of lack of impact," he pointed out.

He mentioned that harnessing the Bay of Fundy is the crown jewel of tidal-energy engineering: "Engineers want the biggest challenge. Greed consumes them." Then he got more animated. "They are going to destroy this ecosystem, then they are going to blame the fishermen . . . and scientists let it happen. Industry-funded science is the bane of our existence. It's socially acceptable to smear fishermen. You can't do that to any religious group, but it's okay to smear fishermen."

About this point, I realized that I was in the cab of a pickup truck with an angry grizzly bear. I glanced at the door handle and considered my options for escape. Darren's cell phone buzzed. He picked up the call and chatted with a fellow fisherman about the Cape Sharp turbines. I was off the hook. "There's no quit in me," Darren said to the fisherman while I listened in. I believed him.

After we drove off the tideflats and onto dry land, Darren described the landscape around us, reflecting on the bigger picture. He pointed out a narrow trail where the deer come down to drink from the stream.

The bush in that flat place is where the male pheasants call out for their mates. He knows this landscape intimately.

Most people think it's a given that the turbines will go in. The fishermen will be pilloried as obstructionists, fish killers, and NIMBYs. The fishermen are not very organized; they are on the defensive and don't have a unified approach. The energy industry's public relations, legal, and political machinery is highly networked, sophisticated, and powerful. They just keep hammering away. Eventually the development will happen. Why else would the power development companies be renting space on the power cable even before they are allowed to plug in? They must be pretty confident this is a done deal.

The energy companies are entrenched deeply into the power structure of Nova Scotia. Dr. Ray Ivany, the president of Acadia University as of this writing, is on the board of directors of Nova Scotia Power. His wife, Laurie Graham, was a journalist until the premier of Nova Scotia, Stephen McNeil, created a special position for her as principal secretary. McNeil is an avid supporter of tidal energy. Much of the funding for Acadia Tidal Energy Institute at Acadia University comes from Nova Scotia Power, FORCE, Nova Scotia Department of Energy, and the Offshore Energy Research Association. It's a tangled web.

Colin Sproul of the Bay of Fundy Inshore Fishermen's Association said in an interview with CBC Radio that last year the harvest of lobster grossed $500 million to the fishermen. The lobstermen are a different class of fishermen in the Bay of Fundy. A lobster permit in Fundy can cost $2 million, and the boats are big and costly—often worth more than $1 million each.

"The reason for our involvement in this is to preserve our way of life and culture," Sproul said. "We will not see it washed away by corporate efficiency and greed." Sproul's association represents 150 lobster and fixed-gear fishermen. "This isn't about money for us—it's about our heritage and culture. It's about money for Cape Sharp and Emera."

Darren Porter says he isn't against tidal power. He believes that in the right conditions, it can reduce our dependence on fossil fuels, but he'd like to see that things are done right. The effect of the turbines on the marine ecosystem needs to be properly assessed before it's too late to turn back—particularly somewhere like the Minas Passage, where high-flow conditions introduce additional unknowns to their operation in a critical ecosystem.

Porter and Sproul have been joined by the organization Clean Ocean Action and more than thirty fishing and environmental organizations supporting efforts to halt the deployment of the turbines until the right tools are available to determine potential damage. In July 2016, the First Nations Mi'kmaq people also outlined their concern with the Cape Sharp Tidal project.

River Teeth

Darren Porter described to me the Bay of Fundy's incoming tide as the "river of life." At low tide, Darren showed me the worn wood stubs of an old weir in the mud. They looked like the skeletal remains of a toothed whale's jawbone emerging from the sediment. He didn't know how old the remnant weir was. He had one of the stubs in the cab of his pickup truck, and he handed it to me in reverence. It was worn and polished smooth.

I thought of David James Duncan's book *River Teeth*. River teeth are the dense knots in a tree where the branches shoot out from the trunk. When a tree has fallen into the river, those knots remain after the rest of the tree has worn away, resembling teeth. It's a metaphor for the "hard, cross-grained whorls of human experience that remain inexplicably lodged in us," after the rest of our narrative has flowed to the sea.

Darren said, "It's not about you or me, it's about the water and the earth. You understand what I'm saying?" I got it. He told me that he doesn't fear standing in the way of an industry known for their aggressive pursuit of profits. He believes that there is a greater power. Individuals have to step up to make a difference in the world and to serve unconditionally. Darren has a bigger vision of the world. His weir fits into a grander picture of his community, Nova Scotian culture, and nature.

There's another story about Minas Passage in Mi'kmaq lore, about how the creator Glooscap saved the world from an evil frog monster who had swallowed all the Earth's water. So when Darren Porter asked, "Whose water is it?" it seemed like an age-old question. But in our times, the availability of technology that potentially alters ecosystems on a massive scale—to metaphorically swallow up all the water—requires us all to come up with answers quickly.[13]

Experiences in a West African Fishery: Counting Lessons

In 2011, a young Yale University graduate named Leslie Roberson went to Ghana to study small-scale fisheries as a part of a postgraduate year. Her goal was to characterize the fishery, the gear used, species caught, participants, value, and so on. She wrote an article for the online journal *Newfound* based on her experience.[14] I interviewed her about her journey.

Prior to her trip to Ghana, Roberson consulted with US National Marine Fisheries Service scientists about collecting data in an appropriate way. She learned about data collection in the Western world and brought copies of forms to fill out—detailing catches of different species, the gear used, effort, and measurements of individuals.

When she arrived in Ghana, Roberson was surprised by the situation she found. The United Nations Food and Agriculture Organization had declared that most West African fish stocks were either fully exploited or on the brink of collapse. She noticed that the catches of the fishermen were so meager she didn't know how they could make a living. They didn't target a certain species, but harvested whatever happened to come up in their nets. She observed that nobody knew the amount of fish the artisanal fishermen were catching or the species, or the extent of illegal and unreported catches by foreign industrial trawlers fishing just offshore. Ghana did not have a fisheries patrol vessel to check on the big trawlers.

There was a fisheries official from Ghana on-site, but his collection of data was sporadic and inaccurate. Species were not recorded. Often the catch was shared with villagers before any recording of the catch was possible. The situation was extremely complex. In the village where she worked, there were actually five distinct communities, each with its own language, economic structure, and fishing method.

Roberson wrote in *Newfound* about one technique she used to get a handle on the situation when interviewing fishermen: she asked them how old they were and how many years they'd been fishing. But some answers made her suspicious, so she switched the order of questions.

"How many years have you been fishing?" she asked one fisherman.

"Ohhh, forty-five years," he said.

"How many years old are you?"

"Me, I am thirty-two," he said.

As she noted, "A contented silence followed."[15]

Roberson described small-scale fisheries of West Africa as difficult to characterize because of the huge range of activities and many participants. Many African governments don't have the infrastructure necessary to monitor such a populous and diverse activity spread over a vast geographic area. She concluded that it's no wonder that "maintaining wild fish stocks at a harvestable level is a colossal challenge that has repeatedly thwarted us."[16]

8: A Dying Fishery?

PUGET SOUND KETA SALMON

"Forty years ago when I started fishing, people said it was a dying industry," said my host Pete Knutson, skipper and owner of the *Njord*. "Now people say the ecosystem is dying. We are doomed."

The deck-level cabin on the *Njord* served as a combination of galley, mess, pilothouse, office, and shop. Sitting on a vinyl-cushioned bench at the table, with an assortment of tools, books, and coffee mugs, Pete gazed out the window through darkness toward the lights of Seattle. A passing ferry droned in the background.

"I wish well-intentioned people wouldn't raise alarms about things they know nothing about. Fisheries go through ebbs and flows. We fishermen change and adapt to them," Knutson said.

On a November night, I observed the harvest and processing of chum salmon on the *Njord* in Elliott Bay, the front yard of Seattle. Fishing for the undervalued chum salmon is one of the adaptations fishermen like Knutson have made to a changing environment. The Seattle skyline was ablaze to accommodate people who were working late in their offices and evening diners in restaurants, while just beyond them in the shadows of the skyscrapers, out on the blackness of the sea, we were catching fish. Chum salmon have not been popular with American consumers and are an underutilized species, but a decline in the other popular species of salmon has turned Knutson to fish for them.

Knutson is a Puget Sound salmon gillnet fisherman who's concerned about the many factors affecting the future of the salmon and the traditional fishermen's way of life. Dr. Knutson is also a professor in the Anthropology Department at Seattle Central College. He was a campus radical at Stanford in the early 1970s, and is still something of a rabble-rouser in the Pacific Northwest fishing community. Knutson started fishing in 1972 because he thought it was an honest way to make a living and be independent. Now he's an activist for his beloved salmon fishery.[1]

Along with his sons, Knutson operates Loki Fish Company, a family business that delivers fish from the sea directly to the consumer. When Pete is on the water, his oldest son, Jonah, is working nearby on his own boat, the *Loki*. Jonah started fishing with Pete when he was twelve or thirteen years old. After college he tried some other work, but came back to fishing. Youngest son Dylan does the marketing and runs the office, occasionally putting a shift in on the boat when needed.

Figure 27. Loki Fish Company's vessel *Loki* setting its nets in Elliott Bay in front of the Seattle skyline. Photographer: Adrienne Higbee. Used with permission from Loki Fish Company.

Figure 28. Skipper Pete Knutson on the F/V *Njord* removing chum salmon from his gillnet, Elliott Bay, Washington. Photographer: K. Bailey.

Dog, Chum, or Keta?

Chum is the most abundant of the wild salmon populations in Puget Sound, yet is underutilized as a fresh-fish product. Each year, about a million chum salmon return from their sojourn in the Pacific Ocean to spawn in rivers that empty into Puget Sound. Whereas most other local salmon species have been in decline over the last century, chum salmon show resilience.

Many people never consider eating the lowly chum salmon because it has been branded as the poor cousin of its tasty relatives: Chinook, coho, and sockeye. Far to the north, Yup'ik fishermen catch chum in the upper Yukon River, but by the time the fish has reached the river to spawn, it has lost fat and become discolored. The Yup'iks call it "dog salmon" because they feed it to their huskies, but they also smoke the fish to eat themselves. It's a name that stuck for many years among Seattle's salmon community.

The chum salmon's reputation isn't completely deserved, because when "keta salmon" (as it is now marketed) is caught fresh from the ocean and processed quickly, the mild taste and flaky texture make it a great eating fish. The new keta salmon is nothing like the snaggle-toothed mushy fish someone's grandfather might have pulled out of the river.

In the ocean, chum salmon are deep-bodied, sleek, and silvery. After three to five years at sea, they move into freshwater to reproduce, and their bodies become a dark olive color with black and red blotchy stripes, sometimes described as a calico pattern. The change is dramatic. The males develop an elongated hooked snout with canine teeth. Soon after mating, the adults die, but their genes are released as sperm and eggs that live on. The spawned-out carcasses deteriorate and become a part of the riverine ecosystem. The released nutrients feed the microbes that form the basis of a food chain that will nurture the river's future generations of salmon.

After fertilization, the eggs take four months to hatch. Then small yolked alevins incubate in the gravel and emerge as one-inch-long fry, which begin to migrate seaward immediately. The cycle of ocean migration begins again, just as it has for thousands of years. The millions of fish and the continuity of death, birth, and life makes one reflect on the meaning of it all. Walking the banks of almost any river in the Pacific Northwest in December, you can smell the dank ammonia of rotting flesh and see hundreds of carcasses strewn ashore, like the bleaching white bones of miniature shipwrecks.

Many scientists worry about the wild Pacific salmon. The most desirable of the wild salmons in Puget Sound, the king, silver, and sockeye, are undergoing declines in population. The causes are the subject of much debate. On the other hand, the chum salmon (along with pink salmon, another undervalued species) are holding their own, perhaps even increasing.

One of the reasons the chum runs are healthy appears to be that they spawn near the mouths of numerous freshwater streams. As a result, they don't depend on the great extent of river systems as they spread inland. Those salmon stocks that run farther into the continent to spawn are those that are faring the worst, because of the habitat degradation that results from mining, agriculture, development, and hydropower activities. That being said, Mother Nature has a way of undermining our sweeping generalizations of how things work by providing outliers. In the Yukon River, the chum salmon migrate almost two thousand miles upstream, and they also migrate deep into the Amur River basin in Siberia.

Gone Fishing: Catching the Salmon

The Loki chum fishery is a story of responsible fishing, careful treatment of the fish product, and sustainable seafood.

"We fish chum and pink because in terms of Puget Sound, these are the healthiest runs," Knutson said. "Right now they are doing particularly well because it seems that ocean conditions are good for these fish. Most of the chum we catch in Puget Sound and all of the pinks are wild spawning populations. The wild coho and Chinook salmon are in deep trouble in Puget Sound because of paving, development, extractive forestry, poor agricultural practices, and generally the lack of strong riparian protections."

According to the Washington Department of Fish and Wildlife, almost all the fifty-five chum salmon stocks in Washington are healthy, while only four of them need protection. The chum salmon fishery in Puget Sound opens twice a week for gillnetters and lasts about a month.

Pete said that the fish run with the tide, so he sets his gear across the tidal current. "You want your net in the water at the change of light," he said. Before he sets his net, Knutson reads the water, a process that he's learned from forty years of fishing experience. Ripples, current, wind, water color, and tides all play a role in the process of deciding where he will set his net.

The 1,800-foot-long monofilament gillnet spooled out from the stern of the boat. Currents billowed the net in a crescent shape. Floats on the top of the net and a weighted line at the bottom keep it suspended vertically from the surface to about one hundred feet deep.

While the net hung in the water, a hefty sea lion cruised up and down the length of it, plucking off salmon. Sea gulls fought over pieces of its meal that floated to the surface. Later we found a disembodied fish head or two in the net.

Pete shrugged and said, "It's like paying the ecosystem a tax."

While the net soaked, Pete was in constant motion. There was a smell of diesel, coffee, and fish in the boat's cabin. Small wavelets slapped on the *Njord*'s hull. Since Pete was sitting in the middle of shipping lanes, he checked the radar frequently. A big container ship skirted around us, cruising at sixteen knots. Pete soldered an electrical part, then peeked back at his net to make sure all was well, prepared soup for dinner, filled out government catch forms. In between, he texted his son fishing on the *Loki*. "Yeah, the action is slower than it was last week," Jonah messaged.

After a couple of hours, the net was hauled back. Pete was multitasking now, steering the boat, controlling the winch, spooling the net, and then stopping to pick salmon out of the mesh. It was a clean catch, with exactly fifty salmon and one dogfish shark. As they came aboard, Knutson carefully slit each salmon in the gills to bleed it and tossed the body into a bin of seawater chilled to thirty degrees Fahrenheit. (Immediate chilling and releasing the blood make a far superior product.)[2] The dogfish was released, and it swam off, no doubt wondering what had just happened.

As a small and independent fisherman it's been hard for Knutson to get a foothold in the fishery. His main competitors are the big purse seiners, which catch the majority of the fish. Even though his catches are clean, environmentalists have demonized the gillnet fishery because of bycatch of other species. There is a constant political effort to shift the gillnetters' share of the catch to the sport fishery or to purse seiners, and this is why Knutson is an activist: to protect his fishery and a way of life for his sons.

It's been a struggle. In the 1990s Knutson fought against two voter initiatives sponsored by recreational fishing groups as well as energy and hydro-dependent industries to ban commercial salmon fishing in the state of Washington. His unlikely allies in the battle were the groups representing the purse seiners. Recreational fishermen blamed commercial fishermen for the declines in king and coho salmon abundance. The energy-hungry aluminum industry and farmers who need irrigation water also wanted to deflect the blame for salmon declines from dam-building activities.

University of Washington professor and part-time fisherman Steve Matthews has said, "But the nets are really a minor part of the problem. What's killing their coho [salmon] is the paving-over of the habitat and the damming-up of the watersheds."[3]

Both initiatives were defeated.

Yet there remains a battle of Knutson's gillnetters with the purse seine fleet over allocation of the resource. The Washington Department of Fish and Wildlife allocates the non-Indian harvest based on time on the water, and there is a constant argument of who gets to fish where and for how long that often ends in the courtroom.

Purse seiners are large boats, typically fifty to seventy feet in length, costing around $3 million each. Because of the efficiency of their method of surrounding the fish in a "purse" and hauling them aboard, they can catch thousands of fish in a matter of minutes. With such high volume, little processing occurs on the boat. The fish are funneled into a hold containing chilled seawater. Other fishermen refer to this type of harvesting as a "dump truck" fishery. Because of the expense of the seine boats and licenses, and the volume of fish handled, there is a tendency for corporations to control the purse seine industry. They also have an advantage in the marketplace, as the large retail chains prefer big-scale deals with major processors over transactions with individual fishermen.

On the other side, gillnet boats are smaller, thirty-six to forty feet long, and cost about a tenth the price of a seiner. Their fish catch is in the tens to hundreds after the nets have soaked in the water for several hours. The salmon are processed on board. The problem for these small-volume fishermen is that their high-quality gillnetted fish compete in the marketplace against the high volume/low cost of the seined fish. And both sectors compete against farmed salmon, which keeps the price low and makes business difficult. There is a glaring price discrepancy between the retail price of fresh sockeye salmon, for example, at $13 to $16 per pound, and the prices paid to fishermen at $.50 to $1 per pound.

It's been tough for Knutson to direct-market his fish; he says that his biggest challenge now is to keep the business going. He's adapted to a changing environment by diversifying and building a vertically stratified company that not only catches the fish, but buys fish, processes it, sells it, and ships it. The direction of his company has evolved into a reticulated path. The in-season costs to transport and process fish are high, and independent fishermen have to self-finance. There are logistic problems of chilling the fish and shipping them. Everything is difficult about being an independent fisherman competing against corporate interests. On the other hand, there is a measure of self-sufficiency and control, and satisfaction of providing food to society and working at sea while integrating into a community of fishermen with a shared culture.

"Has the struggle been worth it?" I asked.

Knutson answered, "I didn't sell myself out at the price of my integrity. I haven't gotten rich, but I'm not morally conflicted."

Back on the boat, the *Njord*'s deckhand Drew Koshar picked a fish out of the bin, lifted it up, and said with respect, "Now that's a beautiful fish," almost as if he were eyeballing a pretty girl at the beach. He picked out another large silver fish and said, "This might be the best salmon we've caught all night." I half expected him to kiss it.

Drew gutted, headed, and scrubbed each fish, and then dropped it into the chilled seawater hold. The time that lapses for fish to travel from the water to the hold is usually less than ten minutes. Later ashore, the fish are graded into levels of quality—depending on their color, their freshness, and the thickness of their belly wall—and they are filleted, smoked, or pickled. Eggs are removed from females and processed into "ikura," a high-quality caviar popular in Asia.

Knutson is right: these guys aren't going to get rich fishing for chum salmon. That night we caught seventy-nine fish. Each headed and gutted fish is worth about $12 to the fishing crew, so that's $950. They get another seventeen pounds of salmon roe, worth another $425. Out of that comes the cost of fuel, maintenance, licensing, food,

Drew's salary, gear, and so on. Their strategy for survival is to vertically structure their company, produce a quality product, and get as much added value from their product as possible. They do much of the maintenance of their boats, they catch the fish, mend their nets, and sell their fish directly to the consumer from the back of their boat, by mail order, or at their stall in local street markets, thereby eliminating the middleman.

Drew also worked in marketing salmon for Loki Fish Company (he has since left Loki, but still works in fish marketing). The company takes advantage of social media and a substantial locavore population in Seattle. Drew told me about a cooking demonstration he gave in nearby Mount Vernon, where he grilled chum fillet with just a sprinkling of garlic salt. Several customers remarked it was the best salmon they'd ever tasted.

But the lack of a consistent quality of the chum sold by others in large grocery chains, and customer acceptance, limits sales of the fresh product. It's an undeserved bad rap. The problem for marketing chum is that many of the chum salmon caught in West Coast fisheries are crushed in the catching process. Seiners chill or quick-freeze their large catches without bleeding or dressing the fish. Then the fish are shipped to a processor, usually in China (thirty-one of forty-six chum salmon processors listed by the Trade Seafood Directory are in China), where they are thawed, processed, and refrozen into individual portions. The product is shipped back to the United States, Asia, or Europe. The quality of the fish available in the supermarket is inconsistent. When the consumer gets a bad product, he isn't going to repeat the mistake. As a consequence, developing a good market for chum salmon has been difficult in the face of the capricious quality and wavering consumer trust of the product in retail outlets.

Two days after my trip on the *Njord*, I bought a fresh keta fillet from Loki's stall in a Seattle farmers market and ran it through an informal taste test of friends and family. At $7.50 per pound for a fresh fillet, this price compared to $29.99 per pound for fresh Chinook and $12.99 per pound for previously frozen sockeye in the supermarket. We tested fresh fillets, frozen vacuum-packed fillets, and smoked keta.

I brushed a marinade (¼ cup balsamic vinegar, ¼ cup olive oil, a dab of mustard, and a dollop of maple syrup) on each fillet and grilled them in a fish basket on the barbecue. When ready, the grilled fish was topped with a heaping portion of fresh mango salsa. I also breaded a piece of the frozen keta, pan-seared both sides in butter, and then finished it off in the oven.

Our unanimous decision: the grilled fresh keta salmon was excellent. It was moist and mild and had big flaky bites. The previously frozen and grilled keta was great, but since they are thinner fillets, one has to be cautious not to overcook them. The pan-fried frozen keta salmon was also delicious. The next day, I made pan-fried salmon patties from the leftovers (mixed with egg, onion, cilantro, cumin, salt, pepper, and diced poblanos), and those were tasty as well. The smoked keta salmon was drier and paler in color than the more familiar sockeye, but was good nevertheless, and is excellent when mixed with cream cheese or served with pasta.

Later I checked in by phone with celebrated Seattle chef Maria Hines of Tilth,[4] an organic restaurant specializing in local foods. "Keta is leaner than sockeye or Chinook, but it is a great salmon nevertheless. It makes sense to use it when you mix this fish with other ingredients," she said.

Good Management Practice

The fishery for chum salmon in Puget Sound is highly regulated and managed cooperatively by the Washington Department of Fish and Wildlife and the Puget Sound tribes. Each year a process called "the North of Falcon" is initiated as a series of evaluations on the status of the salmon stocks in public meetings with federal, state, treaty tribe, and industry representatives to establish the commercial and recreational salmon seasons in the Pacific Northwest.

The fishing season is designed to harvest the number of fish that is surplus to the amount that has been determined necessary for successful spawning in the rivers, called the escapement goal. An escapement goal is made for distinct summer, autumn, and winter runs in each of four major management regions of Puget Sound. Management of the stocks is accomplished by adjusting the amount of fishing time that will be allowed in each area. Specific regional closures and seasonal limitations can be imposed to protect weak stocks.

A preseason forecast of the run size is made based on a number of factors. It's complicated. The preseason forecast of returning salmon is then refined by an in-season test fishery as they reenter Puget Sound waters in October. If the test fishery diverges much from the predicted run, then the duration of the harvest season is adjusted.

Salmon management is complex, balancing needs of the particular fish populations, recreational and commercial fishermen, different species, different gear types, and Indian and non-Indian fishermen. It's a process that's been fine-tuned over the years to operate like a well-oiled machine. But there are hiccups. In spring of 2016 the salmon season was shut down due to an impasse over the tribes' concerns about recreational catches of endangered salmon runs versus the state's support of recreational fishing. Eventually the problem was solved. Maybe it's not always perfect, but it works.

A Future for the Salmon Fishery

Pacific salmon is a legacy fish, a resource that has been utilized and appreciated for generations. The salmon's survival depends on issues that scale from local to global. According to Knutson, the global-scale issues, like climate change, are big problems for the future of the salmon resource. We can fight over local issues like dams and loss of habitat and gain some control over them. But we don't have local control of global issues, and the effects of factors operating at higher-level scales of complexity can be subtle. For example, Knutson noted that in a warming climate, most of the heavy winter precipitation falls as snow in the mountains. But just one warm storm can fall as rain, melting the snow and causing floods at the lower levels, which scour the salmon habitat and subsequently reduce survival of the smolts.

Progress has been made to ameliorate conditions for salmon on the local scale by improving habitat, removing dams, mitigating hydro-development projects, and reducing pollution. But the fight with corporate development over habitat destruction and pollution is continuous.

Henry David Thoreau said in *A Week on the Concord and Merrimack Rivers*, "Still wandering the sea in thy scaly armor to inquire humbly at the mouths of rivers if man has perchance left them free for thee to enter," referring to anadromous fish like salmon. And that is the crux of it. Our choices control the ultimate fate of the salmon. Corporations rally the public to support development projects by waving banners that proclaim more jobs, cheaper power, and increased revenues for local governments. However, the financial gains for the public through development are transient and insignificant compared to the legacy of healthy wild salmon: the salmon provide jobs, wealth, and food in perpetuity if we take care of them.

Big versus Small: The Conundrum of the Pacific Halibut

A seasoned halibut fisherman wrote, "I am tired of the well-funded trawl industry and their lobbyists stepping on the family-run businesses of the halibut industry. Why should my family and I have to suffer while big business is allowed to conduct business as usual?" His letter was directed to the North Pacific Fishery Management Council about an item on their agenda. His complaint was that while the quotas of the directed small-vessel halibut fishery in the Bering Sea were declining, the bycatch limit of halibut by the groundfish trawl fleet hadn't changed and now exceeded the catch of the halibut fishery itself. The issue is a good example of one fishery pitted against another—a small traditional fishery versus a large industrial trawl fleet—and how biology throws more complexity into the situation.

The commercial halibut fishery in the Bering Sea is relatively small, but it is an important nursery area for halibut. A large proportion of the juveniles there eventually migrate to the Gulf of Alaska, where the major fishery occurs. A billion-pound trawl fishery has developed for other groundfish in the Bering Sea, which catches halibut as a bycatch. There has been a limit on halibut bycatch in place for decades, at about seven million pounds.

In recent years halibut harvests have decreased. In 2014 a crossing point occurred when the halibut bycatch exceeded the directed catch of the halibut fleet in the Bering Sea. Because most of the halibut caught in the trawl fishery are juveniles, halibut fishermen have blamed the declines of halibut in other areas on the Bering Sea bycatch of trawlers, and proposed reductions in the allowable bycatch.

In June 2015 the North Pacific Fishery Management Council met to decide on bycatch reductions. The meeting turned into a classic turf battle pitting the rights of one fishery against another. The halibut fishermen enlisted support from recreational charter boat operators and fishermen as well as from their own community of fishermen, complaining about the impact of the trawl fishery on their livelihood. The trawl fishery enlisted the support of fishermen who served as crew on their factory boats and from businesses who claimed that they would lose revenue by potential cutbacks in the trawl fishery.

The council voted to cut bycatch limits by 25 percent. The trawl fishery decried the changes as too much, while the halibut fishermen said it was too little.

Bruce Leaman, who at the time was the director of the International Pacific Halibut Commission (since retired), said that the battle was one of the most complex governance issues he's ever seen in fisheries. The Halibut Commission has the mandate to conserve halibut but lacks authority to manage the trawl bycatch or recreational fishery, while the North Pacific Fishery Management Council has no mandate to conserve halibut, but does have power to manage the bycatch.

STRIKING A BALANCE IN AQUA FARMING

Here and there awareness is growing that man, far from being the overlord of all creation, is himself part of nature, subject to the same cosmic forces that control all other life. Man's future welfare and probably even his survival depend upon his learning to live in harmony, rather than in combat, with these forces.
Rachel Carson, "Essay on the Biological Sciences," 1958

Aquaculture has surpassed traditional capture fisheries as a source of fish and shellfish for human food. As long as the world's human population continues to grow, the demand for seafood will increase even more. Since traditional fisheries have capped out as a food source, the demand for fish will be satisfied by other sources, primarily aquaculture. We can't ignore it. Aquaculture is a potential answer to human food demand if handled with care. In the following chapters positive examples of aquaculture efforts are presented.

There are concerns about ocean farms harming the environment and about the use of pesticides, antibiotics, and hormones to raise fish in intense densities in order to maximize profits; there are consequent effects on the health of the consumer and the rest of the ecosystem. Introduction of diseases to wild populations and the dilution of natural gene pools by escapees are concerns as well. Are there clean, safe, and healthy alternatives?

Here I have chosen to feature two unique aquaculture enterprises. Red abalone is cultured in Monterey Bay with a minimum impact on the ecosystem. The product is high quality and sells itself (chap. 9). Arapaima is grown in Peru and other places. It is not a marine species; however, the culture of arapaima and the harvest of natural populations is a case where much can be learned and applied to marine species. Arapaima is a fascinating fish, and the development to its present status is engrossing and relevant (chap. 10).

9: Mother of Pearl

OCEAN FARMING RED ABALONE IN MONTEREY BAY

The kelp beds of Monterey Bay are giant underwater forests. Holdfasts wrap around stones on the sea bottom to anchor stalks rising up nearly 100 feet (sometimes up to 150 feet) to a canopy of fronds at the surface. Underneath the canopy is an understory of neighboring species. A dense "turf" of matted algae hovers below the canopy like an aquatic groundcover. In and around the kelp forest lives a unique and rich fauna.

Sea otters inhabiting the kelp community are sometimes referred to as a *keystone species*: an animal whose rises and falls have a cascading effect on other species in the ecosystem. The dynamics of sea otters influences kelp-eating animals like sea urchins and abalone. It is a system of feedback and control that has existed for thousands of years. Over the past several hundred years, man entered the scene and has perturbed the abundance of sea otters, abalone, and even kelp, throwing the system out of balance. When an ecosystem strays out of balance—whether due to natural or man-made causes—and the dominant species falls, it opens the door for opportunists to blossom. In this case the opportunists are aqua farmers.

At the end of Fisherman's Wharf #2 in Monterey there is a small building housing the Monterey Fish Company. In the morning chill, a smell of fresh fish and salt hangs in the mist. This is the same wharf John Steinbeck walked while looking for a boat to take him and Ed Ricketts, otherwise known as "Doc," to the Sea of Cortez in 1940. Captain Tony Berry agreed to lease Steinbeck his sardine seiner, the *Western Flyer*, for six weeks.[1] Afterward, Steinbeck and Ricketts coauthored the book *Sea of Cortez*.

Figure 29. Entrance to the Monterey Abalone Company. Co-owners Art Seavey (left) and Trevor Fay (right) pictured. Photographer: K. Bailey.

Just to the north of the fish company, a sign for "Monterey Abalone Company" hangs over a door. The smallish office inside is lined with counters that are cluttered with papers, shells, and instruments. A hatch in the creosote-soaked deck flooring covers a gaping hole. When the hatch is lifted, a wood ladder leads down under the pier. It's dark and dripping under the wharf. There's a smell that blends the essences of fish and other beasts. A pathway of planks leads to the end of the pilings. Pigeons coo overhead, while barks and splashes sound out from the darkness where the planks disappear. As you walk toward the noises, large shapes emerge, and the lugubrious California sea lions, startled, raise their heads in defiance. They grunt, lumber off like overweight men on short crutches, and dive into the water.

Unbeknownst to the tourists sauntering overhead, there is a sea farm of 150,000 abalone under the boardwalk of the wharf. Down here, sturdy mesh cages hang in the sea from a network of beams. There are several hundred abalone per cage, depending on the size of the individuals within. A system of pulleys and ropes is in place to lift the cages out of the water. The enclosures protect the abalone residing within from the marauding sea otters who constantly circle in search of snacks. A worker pulls a cage up and opens the lid; inside are rigid plastic sheets with abalone stuck fast on their surfaces. The capacity of this abalone farm is three hundred thousand animals if there's enough seed to plant them and food for them all.

The abalone are cultured under the wharf to feed man's insatiable desire for the shellfish. It's a complex story, but after the wild abalone population in California was overharvested and collapsed, an opportunity opened for aquaculturists to thrive by supplying the consumer demand.

Abalone and Otters

Abalone, a flattened marine snail, normally creep over rocks and hide in crevasses on the shallow sea bottom along the California coast. They need a hard substrate, and they feed on algae growing over the stones and on kelp that has washed into their habitat. Abalone were a traditional food of the Ohlone and Esselen Indians of central California for thousands of years. The ancient fishermen used the alabaster shells to make fish hooks and ornaments. Based on evidence found in Indian middens on the Channel Islands, abalone were utilized twelve thousand years ago by the first people to populate the Pacific coast.

Seven species of abalone live in coastal California, the most abundant being the red abalone, *Haliotis rufescens*. Red abalone populations cycle out of phase with their major predator in central California, the sea otter. In other words, when otters are abundant, the abalone are scarce.

Otters once numbered one hundred and fifty thousand to three hundred thousand in their range from Russia across the Pacific to Alaska and down coastal California. In 1741, the Russian crew of the Bering Expedition found, killed, and skinned large numbers of otters in the Commander Islands. They sold them in Russia. The pelts be-

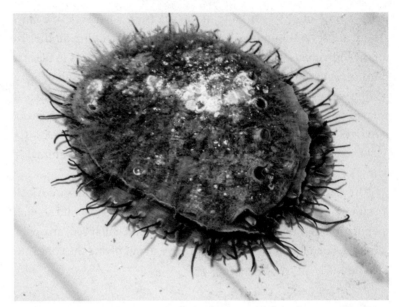

Figure 30. Cultured red abalone from the Monterey Abalone Company. The size is approximately five inches. Photographer: K. Bailey.

came fashionable, and ranked among the world's most prized furs. Then, as the Russian otters became depleted, the "Great Hunt" spread across the Pacific and down the west coast of North America. The harvest of otters continued for over a century.

In the late 1700s, Americans, English, and Russians began trading for pelts with Native Americans. In 1812 the Russians founded Fort Ross in Northern California to facilitate the trading enterprise. Canadians and Americans joined the hunt, trading beads, clothes, and metal implements with the coastal Indians for otter pelts. The British founded Fort Langley near Vancouver in 1827 and traded with the Coast Salish and other tribes (see chap. 4). As the otters became more and more sparse, the value of a pelt increased from $105 to $165 per unit in 1880 to $1,125 in 1903. The incentive to hunt down all the remaining otters was overwhelming. By 1900 there were no more otters in Monterey Bay, and by 1911 there were only one thousand to two thousand otters remaining worldwide. Finally, Russia, Japan, Great Britain, and the United States signed a treaty putting a moratorium on hunting sea otters.

Complex Interconnections

After the otters were overharvested and became sparse, the abalone thrived in the slackened pressure of their natural predators. People along the coast noticed all the meaty snails and began to scavenge the shellfish, especially Chinese immigrants who radiated out from San Francisco for the "abalone rush." The big snails were dried and exported to China. Later, Japanese fishermen joined in, using more sophisticated and efficient methods to reach deeper into abalone habitat with boats and diving suits.

Then the oscillating cycles changed once again. From 1914 to 1970 the protected sea otter population began to increase at 4 to 5 percent per year. They derived from the remnant populations in central California, as well as transplants by California Fish and Game scientists from places where they had recovered into areas in the south where they'd gone extinct. Otters from Alaska were transplanted to Oregon and Washington. In the 1950s the otters reappeared in Monterey Bay. Now the abalone populations were pressured not only by overharvesting, which peaked in the 1950s and '60s, but also by predation from the increasing sea otter population. Harvests of abalone began to fall steeply in the early 1970s, reflecting their declining numbers.

Population Declines

Overharvesting and increasing predation weren't the only problems for abalone. In the 1980s through '90s the red abalone was devastated by "withering foot syndrome," a protozoan disease that affects the abalone's digestive system and causes them to starve. Warm temperatures seemed to promote the disease. Both black and white abalone, with more southerly ranges, were hit even harder and eventually listed as endangered species.

Compounding the abalone's problems, a series of warm El Niño events struck the California coast in the 1980s, disrupting the abalone's food supply of kelp and triggering the proliferation of protozoans causing withering foot syndrome. Finally, in 1997, with populations severely depleted, the abalone fishery closed. The abalone population hasn't recovered. Even now, with commercial fishing banned for nearly twenty years, legal abalone harvesting in California is limited to recreational free divers and shore pickers in the northern portion of the state.

An Empty Niche

When the natural population of abalone began to decrease in the 1970s, their value increased. By the time the fishery closed, the price had escalated from about $2 per pound to $10 per pound, and the delicacy was in great demand.

With a shortage of wild abalone in the sea to satisfy our gastronomic affinities, entrepreneurial scientists began exploring cultivation of the shellfish in the 1960s. First, there was a problem to get them to spawn. Young abalone grow about an inch a year, and they spawn after reaching maturity in three to five years. The problem was that the abalone wouldn't spawn in the laboratory. In the 1970s, a young professor named Dan Morse at the University of California–Santa Barbara discovered that abalone produce hydrogen peroxide when they spawn. (The reason is that hydrogen peroxide stimulates the production of prostaglandin in the abalone, resulting in sexual maturation.) He found that adults could be tricked into spawning on demand by immersing them in a dilute solution of the chemical.

Now culturists could produce a lot of small abalone that swim in the water as plankton, in a developmental stage where each is called a *veliger*. The next problem was that the veligers wouldn't metamorphose and settle on the bottom to complete their life cycle. Once again Morse and his team tackled the problem. They observed that young veligers often settled on crustose red algae. They tested the settlement behavior of veligers in the laboratory using cultures of different types of algae and found that settlement was highest on red algae. Then it was a matter of discovering why the red algae are different.

Morse and his colleagues discovered that a special characteristic of red algae was the production of an amino acid called gamma aminobutyric acid, or GABA, which is a neurotransmitter in animals.

In nature the presence of this chemical guarantees the veligers of settling onto a good habitat. The scientists found that by adding GABA to a culture of veligers, the veligers completed their development and settled on the bottom. More recent studies have shown that veligers also like to settle on the mucous trails left by adult abalone that have passed by.

As a result of these technological advances, by the 1990s commercial hatcheries were producing young abalone to sell to farmers. The farmers start their production with hatchery seed—young abalone about one inch in diameter. They culture them up to five more years, as the abalone grow at a rate of about an inch per year.

Joe Cavanaugh was working for Monterey County looking for ways agriculture could diversify. Then he saw an opportunity: there was a market for abalone and there weren't enough of them from the wild harvest. He became a partner in an abalone farm in Davenport called US Abalone Company.[2] Cavanaugh later separated from US Abalone to form Monterey Abalone Company in 1992. That company is now owned and operated under the fishermen's wharf by Art Seavey and Trevor Fay. Seavey joined the company in 1994, and Fay bought out Cavanaugh some years later.

Seavey got a master's degree in ecology from the University of California in the 1980s during the "blue revolution," when aquaculture was supposed to feed the world's future population. The concept of growing massive amounts of food in the sea stalled shortly afterward, and Seavey moved to Ecuador, where he became co-owner of a shrimp farm. After about ten years there, he moved back to California, where he met and partnered with Cavanaugh at Monterey Abalone.

Co-owner Trevor Fay spends a lot of time on and in the water. Not only does he work the abalone farm, but he motors out into the bay to collect kelp to feed his brood. He is allowed by the state to harvest nearly three tons each week. He also dives to collect marine specimens for a mail-order biomedical supply business and to collect different algae used to make seaweed salads for diners in high-end restaurants.

Trevor's father was a local celebrity in Southern California. Dr.

Rimmon Fay held a PhD in biochemistry from UCLA and owned Pacific Bio-Marine Laboratories, which supplied marine specimens to many research laboratories. One group that he supplied with specimens won the Nobel Prize. It's said that in his day, Rimmon Fay spent more time underwater than any man alive. Over time he turned his efforts to marine conservation and handed the business to Trevor, who later merged his operations with Monterey Abalone.

A challenge that Monterey Abalone encountered as it developed as an enterprise was that sometimes the hatchery production wasn't enough to supply their farm's need for abalone seed. This caused production shortages three to five years later and generated a supply problem for their market. In 2015 the company almost went under due to the shortage. Monterey Abalone decided to team with Moss Landing Marine Laboratories to make their own abalone hatchery. Scientists from the Moss Landing Marine Labs would now be able to experiment with the life cycle and ecology of abalone, and the company would have access to young shellfish for their farm. Fay says that the venture has been expensive, and economically is a break-even deal. On the upside, the availability of young abalone has become more stable. To stabilize their business even more, Seavey and Fay have begun to culture other sea products like sea cucumbers, whelks, spiny lobsters, and red sea urchins.

Monterey Abalone Company sells from sixty thousand to ninety thousand abalone each year. They get $25 per pound live in the shell, depending on size and quantity. In general, the production is limited by the amount of seed available, but in some years the amount of kelp available for feed is the bottleneck. The company grows the abalone for one to five years, and the product ranges from an hors d'oeuvre portion to steak size. The demand for the different sizes varies, which makes planning ahead a difficult task. Still, Fay and Seavey say that the product sells itself. Monterey Abalone doesn't have to do any marketing. The word spreads among chefs, through television cooking programs, popular press, and word of mouth.

Although China (the world's largest abalone producer) and Japan buy a lot of abalone from other California growers, all of Monterey

Abalone's shellfish are sold domestically, and 75 to 80 percent are sold locally on the Monterey Peninsula. Most of the rest are sold in the San Francisco Bay area, with smaller markets in New York and Los Angeles.

Kelp Wars

Relations haven't always been smooth between the environmental community and the Monterey Bay Abalone Company. By the mid-1990s, tourism was bringing in more than $600 million annually to the Monterey Bay area. Abounding with mammals and birds, the kelp beds just off Cannery Row and Pacific Grove were popular destinations for tourists and divers. In the late 1990s Seavey fought what he called the "kelp wars" against a multipronged force of restaurant owners, water-view property owners, environmentalists, and divers protesting his harvest of kelp. Tempers flared.

Things came to a head in 1996 when a commercial kelp harvester from Davenport invaded the bay from their usual cutting area to the north. Using a mechanical kelp cutter like a big underwater combine, they sheared a kelp bed just offshore of Cannery Row. Diners in the waterfront restaurants were outraged. Ironically, some of them were eating the same abalone fed by harvested kelp. Compared to the mechanical harvest, Monterey Abalone's cutting of kelp was small and harvested by hand, but they were lumped in with the big kelp companies.

The Monterey Bay National Marine Sanctuary (MBNMS) and California Department of Fish and Game (CDFG) held hearings to give voice to the debate. The hearings over the kelp harvests became emotional and acrimonious. The anti-harvest contingent once joined hands at a meeting and sang their version of a Pete Seeger song:

Where has all the kelp gone?
Long time passing
Where has all the kelp gone?
Long time ago

Where has all the kelp gone?
Harvesters have picked them everyone
Oh, when will they ever learn?
Oh, when will they ever learn?

Then the protestors handed out cookies.

According to Seavey and Fay, these were frustrating and tense times for their company. Sometimes the protests escalated from verbal to physical encounters. One time a diver aimed his speeding inflatable boat at a Monterey Abalone Company crew that was harvesting kelp by hand from a small boat. The inflatable swerved out of the way at the last minute—spraying the harvesters with water.

Finally, MBNMS sponsored a study which determined that over 220,000 tons of drift kelp were produced in the bay each year. Monterey Abalone only harvests about 250 tons—just over one-tenth of 1 percent of the production—cutting the kelp by hand and carefully rotating the harvest. Monterey Abalone's impact was minimal.

"End of argument," Fay said. In the growing season, the kelp grows over a foot each day. "It's just like cutting the grass in your front yard. That top is going to come right back."

MBNMS determined that there wasn't an ecological problem with the harvest, but rather a conflict among resource users. To mitigate the hard feelings, a panel of scientists was nominated to evaluate the harvest and determine how to minimize the impact on the bay's ecosystem. They recommended to CDFG, which manages the harvest within the sanctuary, to put restrictions on how, when, and where harvests occur. They suggested closing some areas to mechanical harvesting, shutting down some areas to seaweed harvesting altogether, creating a limited entry in some areas, and imposing limits on harvesting from any one kelp bed.

Since then, the controversy has died down. Seavey and Fay say that they don't hear many complaints anymore. Through the ordeal, they learned that educating the public about the small scale of their operation, its local nature, and their attempts to minimize the farm's impact on the environment is critical for gaining acceptance by the community.

Fay says this is a clean and sustainable operation. He believes he is "doing the right thing, the right way, at the right time." His abalone are grown without chemicals or antibiotics added. They are fed local kelp and algae so nutrients aren't added to the bay's system. There is no wasted food, and the sea bottom is swept clean of abalone waste products by strong tides in the bay. Dr. Michael Graham of the Moss Landing Marine Laboratories said the company is "feeding the abalone what they usually eat where they usually eat it. They've pretty much taken all of the negatives out of aquaculture."[3]

Feeding the Couch Potatoes

The company's main problem is still finding enough food of the right quality to optimize the growth of abalone. Farmed abalone must be the animal kingdom's greatest couch potato. In the sea, they are perfectly capable of moving, and they scoot along pretty quickly when threatened by predators. They graze across the ocean bottom in search of food. In culture their movements are limited, and once a week their cages are stuffed with kelp. They just reach one of their two muscular and dexterous feet, called pseudopods, back to grab a chunk of seaweed and pull it into their mouth underneath them. It's like reaching for an endless supply of chips. They hardly move an inch in four years. Finally, after they've been harvested from the plastic sheet where they've resided, a crust of barnacles surrounds the empty shadow where they'd been living.

Finding enough kelp in the winter, its non-growing season, has been a hurdle, and Monterey Abalone has experimented with preserving it during the growing season by drying or salting it. Winter storms can also destroy kelp beds, creating a feast for wild abalone when the loosened fronds wash inshore—but this feast is followed by famine. Some companies have resorted to using artificial feeds. But Fay says that the taste of the abalone on an unnatural diet isn't right, and besides, it isn't practical in the open-mesh cages, as the feed would fall through the mesh to the ocean bottom. (Artificial feeds are more practical in land-based abalone farms and hatcheries.)

Fay says that the abalone grow 25 percent faster if they are fed a diet containing 5 percent red algae. They also taste better and have a more natural color. Monterey Abalone maintains a crop of red algae growing on the wharf's concrete pillars, and also harvests additional algae by diving for it.

Another big problem for sea farming of abalone is the El Niño, which periodically sweeps the California coast. During a strong El Niño event, unusually warm water flows up the coast and the regular process of upwelling stops. It is the upwelling of nutrients from deep water that drives the high productivity of the California coastline and its kelp beds. The high temperatures and low nutrients during an El Niño are a disaster for kelp and the abalone ranchers. Generally the mortality increases from about 2 percent to 12 percent. What happened regarding the El Niño of 2016 was unexpected: it jumped past Monterey, where kelp remained healthy and water temperatures remained normal. But in Northern California, kelp beds were decimated.

Some previous El Niño events had caught Seavey and Fay off guard, so in autumn 2015 when meteorologists began to predict another El Niño—this time a very large one—they began to salt kelp away in barrels, enough for four months.

Best Choice

The story of red abalone on the California coast is complicated. It is the narrative of an interconnected ecosystem: the community of sea otters, kelp beds, abalone, sea urchins, and many other species, clashing with the outside forces of El Niño and man. Everything is tied together.

Our harvest of otters and abalone contributed to the collapse of both species, leaving a vacuum to be filled. Like opportunistic species in a tide pool, the abalone farmers Cavanaugh, Seavey, and Fay saw their chance to provide abalone to a market craving them. Monterey Abalone Company filled a void, and the business has flourished. It does so gracefully and conscientiously, minimizing its impact on the ecosystem. Every aspect of the business is kept as local as possible.

In some other areas of the world, abalone ranches have grown out

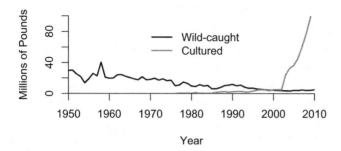

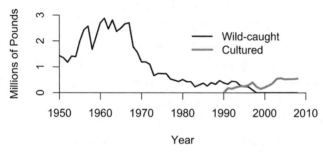

Figure 31. Commercial harvests of wild and cultured abalone. *Top*: Global harvests (data from the United Nations Food and Agriculture Organization). *Bottom*: Harvests of California red abalone (data from California Department of Fish and Wildlife).

of scale. Not only has cultured abalone production filled in for lost wild production, it far exceeds it. Beds on the seafloor are prepared for abalone to be seeded. Predators and competitors are removed, and the ecosystem is altered. The sea bottom is converted into an abalone pasture. Once again, our rapacious capacity to extinguish other animals for accumulation of corporate wealth separates us from the ecosystem.

Seafood Watch said of Chinese abalone sea ranches: "Sea-ranched abalone is rated as 'Avoid' due to the potential significant changes to ocean habitat and the disruption and disturbance to all other forms of marine life that live within the ranch."

On the other hand, abalone farms are a small business in California. There are six abalone farms in production. Three farms are oceanic, and the other three farms are inland, where the abalone are grown in concrete tanks supplied with circulating seawater. When farmed abalone are raised in sea- and land-based enclosures, there is little effect on habitat, no destruction of the natural community structure, and little use of chemicals or antibiotics. As a result of these green practices, the Monterey Bay Aquarium's Seafood Watch lists California abalone as a "Best Choice" seafood.

The 'Namgis and an Ancient Tradition of Ocean Farming

As far back as thousands of years ago, 'Namgis and other coastal tribes in the Pacific Northwest built terraces in the intertidal zone forming so-called clam gardens. The highest concentration of these gardens is in the Broughton Archipelago, British Columbia, just about ten miles from the current 'Namgis reservation around the Nimpkish River and Cormorant Island. There are over 365 terraced areas mapped in the islands. The Broughton Archipelago is now a marine provincial park.

The engineered terraces, which are believed to have been constructed within the last four thousand years, are low rock walls built in the low intertidal zone. Modern ecological studies show that the terraces stabilize the sediments and expand the optimal habitat of the clams. The terraced areas have been shown to support higher clam densities, higher survival rates, and faster growth rates compared to surrounding non-terraced areas.

Cuttyhunk Shellfish Farms

Many artisanal fisheries are family-run businesses like Cuttyhunk Shellfish Farms, owned and operated by Dorothy and Seth Garfield on Cuttyhunk Island, Massachusetts. The company started as an economic venture, but as the Garfield family grew, the farm evolved into a "family business."

The setting for the farm is idyllic. Cuttyhunk Island is on the east side of Buzzards Bay, about an hour's ferry ride from New Bedford. Cuttyhunk Shellfish grows the American oyster, *Crassostrea virginica*. At any one time the Garfields have about two hundred thousand oysters in the water. The oysters are bought as one-inch seed in April and are distributed into collapsible-tiered "lantern" nets. About a year later the oysters are ready to harvest. The annual yield is eighty thousand to one hundred thousand oysters.

Cuttyhunk Island is recognized as a brand of oyster. The taste profile—or *merroir*, as it's known among oyster lovers—is classified as "briny," meaning the flavor is crisp and salty. Cuttyhunk Shellfish is the ultimate sea-to-table operation. An oyster bar on a boat delivers fresh oysters on the half shell to yachts harbored in Cuttyhunk Pond. The Garfields also market their product to fine restaurants in the Northeast.

Hurricanes and storms are a natural threat that keeps the business owners awake at night. In 1991 Hurricane Bob sucked their nets out of the pond into Buzzards Bay. In 2011 tropical storm Irene closed off the inlet to the pond. The town of Gosnold had to cut a new channel because without fresh seawater, the pond becomes depleted of oxygen, killing everything within.

Work on the farm is labor-intensive and presents a physical challenge. According to Dorothy Garfield, working in the family business taught their children the value of hard work and responsibility. As adults, the Garfield children still return to the island in summer to work on the farm. The Garfields hope the tradition of this lifestyle will be passed down to future generations.

10: King of the Amazon

CULTURE AND HARVEST OF ARAPAIMA

I was shopping at a Whole Foods outlet in Monterey when a slab of white meat behind the fish counter caught my eye. The sticker price read less than $12 per pound, compared to over $30 per pound for the king salmon and halibut.

"What's that fish there?" I asked.

"That's paiche [pronounced *pie-chay*], you want a piece? It's really good," the fishmonger said, smiling broadly. He was a slight, friendly man with a Spanish accent—I figured from Central America.

I was looking for something new, and since the halibut and king salmon exceeded my price point, I was curious.

"Is it ocean-caught?" I asked.

"Yes, of course," the man behind the display answered confidently.

I decided to give it a try.

Still wary, when I got home I googled the fish. The fishmonger's story fell apart as I discovered that paiche (*Arapaima gigas*)[1] is farmed, and not even in the ocean. It's grown in the Amazon River basin. I was appalled and considered taking the fish back to the seller in a huff. But it was a thirty-minute drive back to the store in the best of conditions, and now it was rush hour . . . so rather than return the fish, I decided to try it.

In my kitchen I unfolded the brown butcher paper with skepticism and sniffed the fish. So far, so good. I dusted a zesty rub on one side of a fillet as instructed by the seller. I added my own flair of bread crumbs on the other side and then pan-seared it in a mix of olive oil and butter, a couple minutes per side. I garnished the fillet with a dab of Peruvian aji amarillo (a yellow paste made of peppers) and a sprig

Figure 32. Map of the Amazon River basin and the location of Yurimaguas in Peru, where the arapaima is farmed by Amazone.

of cilantro, and then took a white flaky forkful—the fish was mild and tender. Verdict: delicious.

Arapaima is one of a number of new species being raised by a method that might someday be called "organic and sustainable." One of the major issues in marketing arapaima is introducing it to a finicky American public (and it seems that would include educating the fishmongers in local grocery stores).

A Dinosaur Fish

Arapaima is a prehistoric fish whose lineage branched off from its relative, the arowana in Africa, 220 million years ago in the late Triassic (before the Jurassic). The Arapaima genus developed before the Cretaceous, before the Afro–South American drift separating the supercontinent Pangea. Dinosaurs went extinct at the end of the Cretaceous, but not arapaima. Fossils appearing similar to arapaima are found in Miocene deposits. Think of it: dinosaurs probably ate the ancestral arapaima—it must be good for you, right?

A torpedo-shaped fish that flairs toward the rear, the adult arapaima has big diamond-shaped scales that are interleaved like plywood to resist attacks from piranhas. Their color ranges from steel gray to chestnut brown on the back, with distinctive brick-red stripes on the posterior end. The stomach is whitish.

This species ranks as one of the largest freshwater fishes in the world, with the biggest of the beasts stretching to six meters in length and weighing over two hundred kilograms. Besides arapaima and paiche, the fish is also known by a couple other names, including pirarucu, "Amazonian cod," and "king of the Amazon."

I talked with Caroline (pronounced *cay-roh-LEE-nee*) Arantes by phone. She is a biologist who studies arapaima in the Amazon River basin of Brazil. From her photos, she is a petite woman, dwarfed by the size of the fish she studies. Caroline, along with most Brazilian fishermen, calls the arapaima by the name pirarucu, a sound that glides off her tongue like rainforest birdsong.

Caroline said that the arapaima's life cycle harmonizes with the hydrological water cycle of the Amazon basin. Males and females pair off to mate at the end of the dry season, and build a nest in the mud or sandy river bottom at the beginning of the wet season. The females lay around fifty thousand tiny emerald eggs, which the males fertilize. These eggs incubate for three to five days. After hatching, the small fish swim around the male parent for protection. According to some reports, the parent secretes a milky substance containing pheromones from a gland in his head to signal the young fish to stay nearby. The young larvae start off life as alevins, extracting air from water like most other fishes. But as they develop, they begin to breathe air from the surface into a swim bladder that is modified into a pseudo-lung as an adaptation to low oxygen conditions that develop seasonally in the river. Over the next three to six months, the arapaima invade the flooded forest regions to feed on fruits and nuts. They also eat fish and small mammals that venture into the water. Caroline remarked that it is here in the shallow muddy water of the nesting season where they are most vulnerable to fishermen.[2]

Figure 33. Biologist Caroline Arantes with a large arapaima in the Amazon River basin. Used with permission from Caroline Arantes.

Arapaima will defend themselves from humans, leaping from the water like acrobatic salmon and packing a mighty whack. Caroline said one time a large fish jumped out of the murky water and smacked a fisherman, who was assisting her, in the face. As a result, he had to make a visit to the hospital, but was not seriously injured—just stunned.

The fish is also hunted and eaten by caimans and perhaps jaguars in the wild. I watched a recording of a jaguar stalking a large arapaima from a log. The big cat crept slowly toward the fish, visible underneath it in clear water. Finally the jaguar leaped for the fish in the shallow water and after a thrashing struggle, dragged the fish to land.[3] (I waited for Marlin Perkins's sidekick Jim from *Mutual of Omaha's Wild Kingdom* to appear in the scene from this far corner of the world, but it didn't happen.) Caroline said that she'd seen the dramatic footage, and remarked with a laugh that the water was uncharacteristically clear for the Amazon.

Overfished

Arapaima are overfished in the wild. As air breathers, the fish let out a resonating gasp when they surface, making them easy prey for hunters. Local fishermen hunt for arapaima in the rainforest. They stalk the fish with harpoons, listening for the characteristic gasping sound and watching as the fish surface for air. The arapaima take a breath every ten to twenty minutes and are territorial, so the fishermen know about where and when they will surface next. When a fisherman strikes, the fish thrashes violently in the water, until it's dispatched by a clonk on the head with a sturdy club, and rolled over into the canoe.

Before colonization of the Amazon, some indigenous tribes of the river basin may have avoided overharvesting the arapaima through a system of taboos that acted as a form of community-based management. The fish were said to represent an evil spirit, and that someone eating arapaima would be vulnerable to a lightning strike or suffer some other bad fortune, like sores breaking out over their skin. On the other hand, some indigenous people believed that arapaima tongues had medicinal properties that kill parasites. But demand for protein from non-indigenous settlers hunting them for subsistence and as a commodity to sell, and who didn't believe in the taboos, broke down social restrictions.

Salted arapaima was the major source of protein for the early colonial settlers. There was a religious belief of nature's unlimited bounty.

This belief persists, leading some rainforest fishermen to say, "God gave us the fish, then it won't be extinguished."[4] But demand for the fish was greater than nature's ability to produce, and by the mid-1960s arapaima was a rare delicacy. Now they are red-listed as a *probable* threatened species in the wild, lacking detailed information on their population status. They are on an International Union for Conservation of Nature (IUCN) list as a species that is data-deficient, but believed to be overexploited.

The fishermen here live on a diverse range of activities as well, like farming crops and raising livestock. The expansion of the enterprises of logging, ranching, and commercial fishing has led to confrontations with the locals, which has spawned grassroots efforts to take control of local resources.[5] These efforts recognize that a balance between aquatic and terrestrial components of the ecosystem needs to be maintained.

In some Brazilian states, the fishery is co-managed with the community. The organization of the co-management system for harvesting arapaima is based on the networked landscape of floodplain lakes, natural levees, and river channels in order to maintain an ecological balance. Fishermen make counts and are given a quota for harvesting, which is then distributed by the community.

Caroline said that the breathing behavior of arapaima is used by fishermen to estimate how many fish are in a locality. They count the gasps of surfacing fish over a region for twenty minutes. Scientists have validated the accuracy of the method by comparing counts with those from net catches.

Generally 30 to 40 percent of the adult population is considered a safe harvest level. Harvested fish must be greater than one and a half meters in length, and harvesting is allowed from July to November, but not in the reproductive and nursery seasons, when the fish are most vulnerable.

Even though fishing for arapaima is restricted, studies indicate that 77 percent of harvests in the lower Amazon basin are illegal. Local fishermen conduct "surveillance" of their areas to keep out "invaders" from outside the community. But the bigger motorized boats outrun the local dugout canoes. Hearing Caroline talk about how local fishermen along the river participate in the monitoring and management, the vast scale of the area, and the shortage of resources makes me realize how complex and logistically difficult it is to manage small-scale fisheries in much of the world. The landscape of the floodplain is complicated and remote, and changes dramatically from season to season. Fishermen make deliveries up and down the river, there is little reporting, and few eyes are watching.

Caroline explained that overfishing is still the major concern. But there are other threats to the sustainability of the arapaima population. The fish are dependent on the forests for food and as nurseries for the offspring. "Fifty percent of the area in the floodplain of the lower Amazon has been deforested in the last forty years," she said. Grazing by cattle and water buffalo on grasses along the shoreline has degraded the quality of the nearshore habitat. The water cycle is critical to the arapaima's life in the river and floodplain; dams, construction, and climate change have had an impact on the flooding and drying of the forest. Caroline noted that it seems with the changing climate, periods of drought are getting longer and flooding is becoming stronger. These factors affect the recruitment of young fishes into the population.

Culturing Arapaima

With the arapaima threatened in the lower reaches of the Amazon, people have begun to grow the fish in the upper reaches. In 2008, a Peruvian regional government project to culture arapaima was abandoned and 142 adult fish from the project were acquired by a Peruvian start-up company called Amazone. They moved the fish to specially created ponds in the Amazon rainforest, near Yurimaguas, where the fish have thrived.

Yurimaguas is located in the northeastern corner of Peru in the region of Loreto. At the confluence of the Huallaga and Paranapura Rivers, it is an important port for export of forest products, legal and illegal. The area around the town is farmed for sugarcane, bananas, cotton, tobacco, and manioc.

According to Isaac Gherson, the manager of the Amazone proj-

Figure 34. Aerial view of the company Amazone's arapaima farm, near Yurimaguas in Peru. Used with permission from Paiche-Amazone.

ect, arapaima breed naturally in the ponds. In cultivating arapaima, the offspring are removed from the adults and fed pellets made from Peruvian anchovies, soy, and grains, without other animal by-products.[6] They are grown at low densities, and the ponds are continuously flushed with fresh clean water from the Amazon. If you are worried about Fukushima radiation in your fish, one can't get much farther away from there on this planet than the Amazon rainforest.

Arapaima grows faster than other farmed fish. After eighteen months when they are ready to harvest, the fish are about 1.3 meters long and weigh ten to twelve kilograms. They are processed in a clean, modern facility and super-chilled immediately. The dressed-out yield of product is about 60 percent. Production of arapaima for export by Amazone has increased dramatically, from five tonnes in 2010 to 250 tonnes in 2013.

Environmental Concerns

I had a number of questions about the process of cultivating arapaima. One of the investors in Amazone is the Hochschild Group. They are a mining conglomerate and have been involved in bauxite mining in the Peruvian rainforest. Bauxite mining scrapes soil from the surface, leaving shallow pits, which raised my suspicions. Could they be using scars in the earth created by bauxite mines as rearing ponds? Are there toxic tailings left behind? I read a report in *Time* magazine that the ponds were left over from a cement plant. I fired off an email to Amazone's Gherson, who assured me that is not the case. The bauxite mines are more than 160 kilometers from the arapaima ponds, which, he said, were dug specifically for culturing the fish.

Gherson had told me that antibiotics and hormones aren't used in growing arapaima. But a week later I woke in the middle of the night. We hadn't mentioned pesticides. My mind churned about the thought of poisons used to control mosquitoes in the ponds. This would be a deal-killer. In the wee hours of the morning I fired off another email to Gherson, who by now must have been concerned about my mental health. Gherson reassured me that no pesticides are used in the ponds. (I can laugh now about my anxiety; however, I think it's reasonable to be wary about the food industry, what we purchase from them, and then put in our bodies.)

Gherson said that he has concerns about pollution, deforestation, and climate change in the future of arapaima. However, according to him, the Peruvian government is working hard to control illegal mining operations that pollute the water. Then there is illegal logging, which affects not only the habitat of wild arapaima but also the water quality of his culturing operations. Warming water temperatures due to climate change are also a concern, and even though arapaima are tropical, they have limits to their thermal tolerance.

The arapaima exported by Amazone and marketed as paiche are registered and certified as farmed fish, and their harvest does not affect wild populations. But Amazone scientists are also interested in the health of wild populations. Gustavo Sakata from Amazone approached the WWF (World Wildlife Fund) about introducing arapaima to natural lagoons where the stock had been depleted. The hope was that in addition to restoring the natural balance, it would benefit native fishers who supply local markets and might eventually develop export markets. Sakata explained to Aldo Soto, a rainforest biologist formerly with WWF, that this was a part of their company's policy of environmental and social responsibility. WWF biologists determined that the introductions should be from the same species/subspecies of the natural population so that the natural genetic diversity of local fish is not compromised.

Amazone reintroduced some individuals into a lake in the Kichwa community called Campo Verde, much to the delight of the indigenous community. Other restocking projects have also been successful—for example, in Imiria Lake in Ucayali. But biologist Soto noted that the river systems in Loreto and in the lower Marañón basin are highly connected, and it's critical to carefully monitor the genetic basis of the stocking program so that locally adapted subpopulations are not affected by mixing, thereby losing biodiversity.

Further downstream in the central basin of the Amazon, Caroline noted that as long as fishing levels are controlled and there is no illegal fishing, arapaima don't need to be stocked because of the population's high growth potential. Natural populations can grow at 25 percent per year, so depleted stocks can recover quickly, provided that there is still a source of reproducing fish and fishing mortality is controlled.

There is controversy about whether arapaima belong in the upper reaches of the Peruvian Amazon. One report indicates that arapaima were artificially introduced to the upper reaches of the Amazon basin of Peru in the 1960s; the fish is indigenous to the middle and lower reaches of the Amazon. Furthering the controversy, in the 1970s, the species was introduced to the Bolivian Amazon, where it didn't exist naturally, but now reigns as the most important fishery.

Marketing Arapaima: Keeping It "Green"

Amazone's efforts have made arapaima visible internationally and have opened global markets. These efforts have improved the market conditions for arapaima, and as a consequence, local fishermen may someday be able to export harvests from managed lakes.

The fish has become popular with chefs but has not been flying out of the counters in US retail markets. Consumers are wary of new products.

"First of all, there must be a niche market that values the product for its properties and is willing to pay for it. In the US there is a gourmet market that wants this kind of farmed fish," Gherson said.

There also is demand for the fish in Asia and Europe, making it competitive and harder to find in American markets.

Arapaima has become popular as an ingredient in trendy Peruvian restaurants that are springing up in Boston, Los Angeles, Miami, and other cities. Peruvian cuisine is unique, sometimes described as a fusion of South American and Asian cooking. I talked about paiche as an ingredient with Ricardo Zarate, who was the chef/owner of three restaurants, Mo-Chica, Picca, and Paiche, in Los Angeles.[7] "I love this fish," he said. "It's an amazing fish, like a mix between black cod and monkfish." His favorite presentation is battered and fried in butter, but Chef Zarate added that paiche can be served as ceviche (thinly sliced or carpaccio style), baked, or grilled (be careful not to overcook it, he warned).

Amazone's product is imported to the United States by ArtisanFish, a company that markets healthy and sustainable fish. Then Whole Foods, which uses its own green-buying fish brokers to source responsibly farmed or caught seafood, retails the fish to consumers. The operation is audited by Institute for Marketecology, a third-party certification company.

The arapaima has 20 grams of protein per 100-gram serving. This is comparable to a beef fillet or chicken breast, but with a lot less fat (1 gram) and fewer calories (89). Farmed arapaima is high in omega-3 oils[8] with a favorable ratio to omega-6s. Compare that to another "white meat," the pork chop, at 19.3 grams of protein, 20 grams of fat, poor omega ratio (depending somewhat on what the hogs are fed), and 260 calories. Thanks—but I think I'll take the arapaima. Arapaima (at least those grown by Amazone) is raised without antibiotics, vaccines, or hormones. The product is free of mercury and heavy metals.

Ethical Dilemma?

When I first learned about cultivated arapaima, I was dismayed at the thought of commercial development in the Amazon rainforest. On the other hand, it's easy for us to say that any project in the rainforest is undesirable while we live in modern comfort. In doing so, we deny the indigenous people and Amazon residents who live in the rainforest the opportunity to control their own fate. Can Westerners preserve the Amazon and the Amazonians as a living museum and deny them access to our own lifestyle? Would we switch places? Maybe a reasonable approach is to encourage and support low-impact and locally owned development that is sustainable to the health of the rainforest and its inhabitants.

For fishermen fishing in the river system, community-based management is a tool that has been developed to help artisanal fishermen co-manage wild stocks of arapaima, bringing benefits to many people. The scheme used in some regions of the lower Amazon has been effective for recovering overexploited arapaima populations. The method has improved yields, generated profits for fishermen and their families, and enhanced community organization.

Some conservationists think that wild arapaima harvests should be reserved for local or national consumption, while farming is expanded for export products. As the global market demand for arapaima grows, more companies will get involved in arapaima farming. Not all of them will be socially responsible. Companies that cultivate arapaima are springing up in Peru and Brazil as well. Brazil intends to increase the production of arapaima and tilapia to more than two million tonnes annually, with most of the growth in arapaima production. There are reports of the fish being raised in regions outside its natural habitat, like Chile, Malaysia, and even Florida. One imagines China will jump on the bandwagon at an overwhelming scale. How can we protect ourselves from unscrupulous practices, conserve the Amazon rainforest, and encourage responsible farmers? One way is by consumer education and demand.

The recent resurgence of organic farming in the United States and Europe that has been fueled by consumer revolt against industrial agriculture and the use of pesticides, herbicides, and hormones in food products can serve as an example to aquaculturists. There is consumer demand for organic produce and meats, such that they are now stocked in major retail outlets. There is likewise a market for cultured fish products from certified reputable sources—those that guarantee that the product is farmed in a socially and environmentally responsible manner.

In order to supply produce to retail chains, the small organic farms had to grow large. Now some are controlled by much larger corporations. Likewise, Amazone is not an "artisanal" aquaculture venture like Monterey Bay Abalone. It is a corporate venture. This example points out the difficulty in maintaining small fishery businesses in large national and global marketplaces. Big retailers want a continuous supply of product. In the case of Whole Foods, they had to put resources into marketing paiche before consumers would buy it. In order to satisfy Whole Foods, Amazone had to grow bigger.

Food safety, food quality, sustainability, and adherence to human rights and international labor standards are issues in food production. Producers, middlemen, and retailers sometimes misrepresent their product, even when selling to companies as fastidious as Whole Foods.[9] One ongoing concern is the level of toxic organic and inorganic substances in the fish pellets used in aquaculture. In particular, the use of organic antioxidants—required by international shipping regulations to prevent combustion—has been brought up as a concern. There is no substitute for vigilance by retailers, consumer action groups, and government agencies to protect the safety and security of our fisheries resources.

Fishing Cooperatively with Brazilian Dolphins

Man has used dogs and falcons to assist hunting on land for thousands of years. There are parallels in aquatic systems. In Japan, China, Peru, and Greece, fishermen trained cormorants to catch fish since before AD 1000. They tied a snare around the bird's neck to keep it from swallowing the larger fish that the humans wanted to retrieve. Now, fishing with birds is a show for tourists. For centuries, fishermen in Europe, Asia, and Africa domesticated otters, trained as pups, to help them catch fish. The otters are still used in Bangladesh, where they chase fish into waiting nets. But as far as I can tell, the cooperative employment of dolphins by fishermen in Brazil is the only use of completely wild animals to assist in fishing.

There are two regions in southeastern Brazil where wild dolphins and humans cooperate to catch fish. On a beach called Tesoura in the city of Laguna, cooperative fishing with dolphins has been going on for more than 120 years, and perhaps as far back as the sixteenth century, by fishermen who immigrated to Brazil from the Azores. The dolphins give specific signals with their beaks, showing the fishermen where to cast their nets for mullets and anchovy. Out of fifty-five resident dolphins in the lagoon, twenty-one cooperate with the fishermen. The cooperating dolphins are called *botos-bons*, or "good dolphins." It's believed that the fishermen's nets disrupt schools of fish, scattering them, and in the chaos the dolphins are more successful predators. On the other side, the fishermen who cooperate with the dolphins have a higher catch rate and catch bigger fish than fishermen who don't cooperate with the marine mammals.

To the south 160 kilometers, at Barra do Imbé/Tramandaí, the cooperative fishing by men and dolphins plays out similarly. It's said that the botos-bons recognize certain men in the water and guide the fish toward them. The fishermen here are concerned that tourists using sport watercraft—which scare the dolphins away—will disrupt the fishery.

Veta la Palma and Other Innovative Aqua Farms

Ocean ranching and especially "green" methods are springing up all over the world.

Veta la Palma is a commercial fish farm in southern Spain near Seville. It produces 1,200 tonnes of sea bass, bream, mullet, sole, meagre, eel, and shrimp each year. The farm calls itself a natural park, with two hundred species of birds utilizing the pond ecosystem. Previously the area had been a degraded wetland, drained of water to grow rice. The fields have been reflooded to form the ponds. Bream, bass, meagre, and sole are reared in a hatchery and released in the ponds to grow. Eel and mullet enter the ponds naturally from the estuary. All are grown at a low density, and no antibiotics are used in the ponds. There is a strong tidal exchange of water in the ponds that keeps them well oxygenated. Twenty percent of the fish are eaten by birds, which is an "ecosystem tax." Fish in the ponds are not fed a commercial diet, but eat shrimp that thrive naturally in the ponds.

There are many other examples of innovative aquaculture techniques. Carlsbad Aquafarm, in operation since 1990, grows mussels, oysters, abalone, scallops, and ogo (a seaweed) on Agua Hedionda, a shared property with a natural gas power plant in Southern California. Earth Ocean Farms, an aquaculture operation off the coast of La Paz in Baja California, Mexico, grows fish in giant offshore cages. Catalina Sea Ranch grows mussels that are attached to a system of ropes suspended twenty feet below the surface over the San Pedro shelf, six miles off the coast of Southern California. They are planning to have sea cucumbers below the mussels to feed on detritus from the shellfish. GreenWave is a three-dimensional ocean farm being tested at Thimble Island Ocean Farm; it is billed as an undersea garden to grow seaweeds and mussels on ropes suspended in the water, and oysters and clams in cages below.

11: Evolving Solutions

We stand now where two roads diverge. But unlike the roads in Robert Frost's familiar poem, they are not equally fair. The road we have long been traveling is deceptively easy, a smooth superhighway on which we progress with great speed, but at its end lies disaster. The other fork of the road—the one less traveled by—offers our last, our only chance to reach a destination that assures the preservation of the earth.
Rachel Carson, *Silent Spring*

In his book *Words of the Lagoon*, Robert Johannes tells the story of fishermen in Palau, an island country in Micronesia. Seafood was once abundant there. The Palauan fishermen never had trouble finding enough fish to satisfy their own and their village's needs. The fishermen gave away the fish they didn't eat to other villagers.

The village controlled access to fishing on the reefs adjacent to them, similar to the old Sámi and Coast Salish traditions. "In the old days we would take the fish we needed to eat . . . and let the rest go," said one old fisherman. They lived in a state of "subsistence affluence." The Palauan value system was not based on "elevating one's self through controlling the production of others, as is done in most industrial societies."

After Japan colonized Palau in the 1920s, the fishermen began to sell their fish to obtain attractive and exotic goods offered by the Japanese. The fishermen bought nets and motorized boats with the money, allowing them to catch more fish to sell in order to obtain more goods. They fished harder to harvest more fish and visited more distant areas of the reef to find them. Over the years, the fish abundance dropped.

The fishermen bought even bigger boats to catch the vanishing fish, but to do that they had to borrow money. They had to sell all their fish in order to pay off their loans. They stopped giving them away in the village; instead, they sold them to the outsiders and to other villagers. Now the people in the village had to work for the money to buy their food. The traditional social and economic fabric of the community deteriorated.

Pretty soon, there were not enough fish over the reefs for the fishermen to make payments on their loans, so the village sold their customary access rights to the fishing grounds. The people in the village began to eat imported fish in cans.

The narrative of the Palau fishermen sounds like that of the historical Sámi and Coast Salish fishermen. Like many other fishermen I listened to as they told their stories, they all have experienced a common theme: the clash of an established system with a new order put forth by colonists, developers, and industrialists. The new order stakes a claim on the fish, the market, the shorelines, or the water.

The artisanal fishermen and small-scale aquaculturists featured in this book share a love of the sea, the freedom of independence, and passion for their work. Each of them confronts a wide range of problems in a struggle to survive. Their stories tell of encroaching development and habitat destruction, competition with industrial fishing, declining fish stocks, adverse climate, government restrictions, and loss of fishing rights. The fishermen's need for healthy ecosystems to support healthy fish stocks conflicts with use of the water by energy and extraction industries, or the "find, fish, and run" harvesting strategy of some multinational fishing companies. In many cases, the artisanal fishermen have adapted to changes in their environment, and in others they—alongside indigenous fishermen—lead the charge to save their fisheries and balance social justice.

"Who Fishes, Matters"

Niaz Dorry is the coordinating director of the Northwest Atlantic Marine Alliance (NAMA). NAMA was founded in 1998 as a nonprofit organization "dedicated to pursuing community-based management to achieve its purpose of restoring and enhancing an enduring marine system supporting a healthy diversity and an abundance of marine life and human uses through a self-organizing and self-governing organization." They aim to educate the public on the state of fisheries, the issues of food security, and what can be done about it.

I arranged to meet Dorry in a coffee shop in Seattle. On a sunny summer day sitting in the café's patio, we talked about fisheries as naturally as the pair at the table next to us talked about backyard gardening. Before joining NAMA she was a Greenpeace activist. Now Dorry advocates for community-based fisheries.[1]

Not only is Dorry involved in the sustainability of fisheries but in the broader issues of social justice. She pointed out that Seafood Watch, a program of the Monterey Bay Aquarium to rate fisheries, is helpful for conservation of fisheries, but doesn't go far enough. It protects the resource, but not the humans involved in the industry. According to Dorry, Seafood Watch responded to her complaint by saying, "Our scientific assessments don't incorporate social issues."

The biggest issue in fisheries right now, in Dorry's view, is privatization. Privatization through catch share programs is leading to industrial consolidation and the loss of small-scale fishing. Bigger companies are more efficient and cut some costs, leading to increased profit—but with a loss of jobs and community infrastructure; that's the economy of scale. But, Dorry says, "Who fishes, matters."

NAMA has documented the process of privatization in New England groundfish fisheries, their consolidation, and the selling or leasing of quota. Quota is too expensive for small-scale fishermen to buy. In 2012, Atlantic cod quota out of New Bedford cost $2.48 per pound to lease. In general the lease fee is 85 percent of the fish's value, thus the fishermen get only 15 percent, minus expenses. Dorry says the only way to survive is to get big or get out.

One of the advantages of small-scale fisheries, as Dorry points out, is fleet diversity, or the ability to switch target species as they become abundant or depleted. The adaptability of the fleet is more like natural predators feeding in an changing ecosystem, rather than concentrating effort on a single species. But with catch share programs the small fisherman can't be invested in so many fisheries because of the expense of obtaining various quotas. Catch shares are determined by historical catch rates, and the biggest companies capture most of the share.

"Codfathers" and Sea Lords

The story of Carlos Rafael illustrates one of the problems with catch shares and the current system of management. Rafael, also known as the "codfather" of New Bedford, Massachusetts, controls a dominant share of the New England groundfish quota, including one-fifth of the cod quota.

Rafael grew up in the Azores, where he attended school in a monastery. He immigrated to the United States as a teenager in 1968. He worked as a fish cutter on the waterfront, then got into seafood distribution and supply. In 1981 Rafael bought his first fishing boat; his fleet eventually grew to forty-six boats. He could have struck a pose as the poster boy for the American dream.

Except for one problem. Rafael has been described as "a cutthroat capitalist who is perpetually at war with someone: regulators, competitors, environmentalists. He battles, forever with an eye on his profit margin."[2] In the 1980s, he spent four months in jail for federal tax evasion. In the 1990s, he was charged with price-fixing his fish, but acquitted.

Rafael is the biggest fishing boat owner in the United States. He dismisses his competition—describing the smaller New England fishermen who compete with him, he once said, "They are like mosquitoes on the balls of an elephant. Biting, biting, biting, until finally [the government] is going to say fuck off, we got to do something."[3]

On February 26, 2016, Rafael was arrested by federal agents on twenty-seven counts, including conspiracy, false entries, and cash smuggling. It was disclosed that "Rafael is alleged to have falsely reported the species of more than 815,000 pounds of fish to the National Oceanic and Atmospheric Administration (NOAA) between 2012 and January of this year [2016], according to the Office of U.S. Attorney Carmen Ortiz." The report continues: "Prosecutors allege that Rafael lied to the government for years about the amount and kinds of fish caught by his fleet, in order to evade federal fishing quotas, then sold the illegally caught fish to a New York City buyer, for cash payments of hundreds of thousands of dollars."[4]

About the kinds of fish he caught, Rafael allegedly said to an undercover agent, "I can call them haddock. This year I'll have 15 million pounds of haddock [quota]. So I can call any son of a bitch haddock if the bastards are not there. I rename them. Even when they're there, I disappear them. I could never catch 15 million. It's impossible."[5]

Violations of regulations in catch share programs by mislabeling fish, overharvesting the catch quota, or breaking other rules are not uncommon. For example, in 2015 about 17 percent of vessels fishing for Pacific halibut were cited by the US Coast Guard.[6] Some violations can be on a much larger scale than the misdeeds of individual fishermen. American Seafoods, a company with a large share of the Alaska pollock quota, was cited by NOAA for underreporting their catches in multiple years and on several ships.[7] A hearing scheduled in 2014 was canceled after reaching a negotiated settlement of $1.76 million—a paltry sum to a fishery worth a billion dollars. The company representative said the settlement highlighted "the complexities of operating, maintaining, calibrating and testing sensitive equipment like flow scales in the demanding environment of an at-sea processor."[8]

The economists who designed catch share programs expect fishermen to behave like farmers, who take care of the soil and set aside seeds for the next season; they believe fishermen should behave rationally and conserve the resource. But in reality, the nature of fishing is different from farming. When a fisherman casts his net into the sea, he doesn't know what he's going to catch. The fish aren't fixed to a piece of land. Gaspar Catanzaro of the Monterey Fish Company said, "You know, being a fisherman is like going to Las Vegas, and you put the money in the slot machine. Either your money is going to come back out or you're getting nothing."[9] Like with gamblers, the element of risk and immediate gratification influence the fisherman more than other resource users.

There are examples where catch shares are reported to be working, such as the Gulf of Mexico red snapper.[10] EDF says that the IFQ program there has rescued a depleted fishery. Others say it isn't the catch share program that has rebuilt the fishery, but stricter regulations and monitoring. Now, over half of the harvest of red snapper is controlled by investors who never even fish, sometimes called sea lords. For them the fish are a commodity. One individual is said to control 30 percent of the red snapper harvest by leasing quota. Another big quota holder reportedly branded opponents of the catch share program as "communists." He is referring to the independent fishermen who want to openly compete for the harvest, not fish as tenant sharecroppers. IFQ supporters mention other IFQ successes in Australia and New Zealand, but many of those fisheries are dominated by investors, and not fishermen. From an economic standpoint, perhaps these are successful fisheries; somebody is making money, but it isn't the fishermen.

According to many fishermen, the catch share programs aren't working for them.[11] In 1994 an IFQ system for the US halibut fishery was formulated, and quota shares were granted. Nearly five thousand fishermen on about 3,500 boats were given quotas. Fishermen were allowed to sell their quotas, which they did in an avalanche. In spite of rules to discourage it, the fishery consolidated. Now there are about one-third the number of boats in the fishery. Lots of ex-halibut fishermen got wealthy by selling their quotas. Meanwhile, the retail price of halibut for consumers has increased dramatically, becoming a luxury food for most people. In 2016, the price fishermen received at the dock was $6 to $7 per pound. The price to buy quota shares was $50 to $70 per pound. That's a major investment. Only 30 percent of share buyers are younger than forty; the major age group of buyers is fifty to fifty-nine years.[12] In addition, there has been a problem of "mailbox fishermen," or "armchair fishermen," who lease quota for 50 to 85 percent of the landed value. Amendments to the halibut IFQ program have attempted to short-circuit these problems—but enforcement is lax. In Canada the halibut quotas of many small-scale fishermen were bought out. By 2006 over half of the total quota was owned and then leased out by armchair fishermen. The halibut leasing arrangement has been compared to a sharecropper system. In 2015 the fishermen who leased quota received only 15 percent of the landed value of the fish they caught—a narrow margin to pay crew and boat operating costs. The lessors pocketed the other 85 percent of the value of the harvest.[13]

Perhaps catch shares are a reasonable way to limit the number of boats and conserve fish populations in the sea, but managers are still working out the kinks. The big issues—such as the concentration of catch shares, inheritable shares, and armchair fishermen—could be fixed if there was the political will to do so.

Many think that consolidation of fisheries already has gone too far. A study by Osterblom et al. (2015) reported that thirteen transnational corporations, operating a network of subsidiaries, control 11 to 16 percent of the global marine harvest, and 19 to 40 percent of the largest and most valuable stocks.

Community-Based Fisheries

Port Orford is a small picturesque town on the Oregon coast that has been hit hard by the slowdowns in the timber and fishing industries. In the summer of 2015, I met with Leesa Cobb, the executive director of the Port Orford Ocean Resource Team. Cobb turned toward a wall in her office, and with a dismissive wave of her hand at some plaques hanging there, she said, "It hurts me to look at them." Her organization has been given awards from NOAA and the state of Oregon for their work on community-based fishing. She said, "They acknowledged our work but failed to recognize what it was we were really attempting to do."

The Port Orford Ocean Resource Team (POORT), like NAMA, is part of a growing network of community-based fisheries ventures. They are activists with an interest in linking consumer choice in the seafood they buy and the sustainable fishing practice of the fishermen whose products they support. Each of them sees that management of fisheries is coast-wide, and not locally based as they believe it should be. Current fisheries management is top-down rather than bottom-up. Their goal is to empower fishing communities in the process of managing sustainable fisheries.

Of the 1,200 people living in Port Orford, about one in ten depend on fishing. There are forty-five fishing boats in the Port Orford coastal fishery, all forty feet long or smaller. They fish with hook and line for a variety of fishes, including black cod, rockfish, tuna, salmon, and halibut; and they fish for crab with pots. It's the ability to switch from one fish to another, the diversity in the species they harvest, that provides these small-scale fishermen of Port Orford the chance to make a living from the sea.

There are problems in this fishing and logging community. The fisheries bring $4 million into the local economy, but the child poverty rate in the Port Orford school district is high, according to the US Census Bureau—about 55 percent, equal to the poverty level of inner-city Detroit. The poverty links up with other social problems, like depression, drugs, alcohol, and homelessness.

The fishermen of Port Orford are graying. Their average age is fifty-three. It's expensive for younger fishermen to get a foothold in the fishery. A salmon permit sells for about $5,000, and a nearshore groundfish fish permit goes for about $30,000. Crab permits sell for anywhere from $60,000 to $300,000. A seaworthy boat of twenty-five to forty-five feet will cost $30,000 to $100,000 and up. The advantage Port Orford does have for the younger fishermen is that the small boats used here are relatively affordable compared to seiners and trawlers, which cost hundreds of thousands to millions of dollars.

The original goals of POORT were to conserve habitat and to protect the local area from overfishing. They conceptualized that local marines reserves would be established under the direction of local fishermen and that the local community would be participating in self-managing and self-policing the fishery. The local fishing grounds and upland watersheds would be a community stewardship area.

Cobb argued that as of summer 2014, there was nothing in NOAA's West Coast groundfish catch share program that benefits small-scale fisheries. She said that small-scale fishermen and local communities don't have a voice in the process. Local fishermen are marginalized in the management process since they don't have a quota like the big trawl fleet. Ironically, the concept for POORT started with a grant from EDF, the major proponent of catch share programs.

Some thirteen years after starting their work in Port Orford, Cobb thinks that their science initiatives, based on finding out more about the local ecosystem, as well as their marketing efforts have been very successful. Port Orford Sustainable Seafood launched a community-supported fishery in 2012 that markets locally caught fish directly to consumers from Ashland to Portland. Port Orford fish are recognized as a brand. However, their endeavor of community-based management has been less successful because they haven't been able to get attention from the Pacific Fishery Management Council, and fishermen have resisted establishing marine reserves.

Real Good Fish

Commercial fishing is important in the central California community of Monterey, where it has generated over $70 million in earnings at the dock between 1990 and 2011. Along with recreational fishing, commercial fishing supports over 750 jobs and brings in millions of dollars in tourism-related spending. In 2011 Monterey ranked thirtieth in the nation in fisheries landings out of 1,500 ports. Some twenty-five million pounds of seafood are landed in Monterey annually.

A community-supported fishery similar to that in Port Orford is centered in Monterey Bay. Alan Lovewell, along with partner Oren Frey, founded Local Catch Monterey Bay in 2012. The plan was partly modeled on the first community fishing operation off the West Coast in San Francisco. In 2015 Lovewell continued the venture alone as CEO of Real Good Fish (RGF), and he expanded his area of operation to the San Francisco Bay area. The company supplies fresh seafood directly from local fishermen to subscribers on a weekly or biweekly basis, a sea-to-table venture.

I met with Lovewell over a beer in downtown Monterey. He's a busy young man, with a full platter of seafood business—meeting fishermen, delivering fish, and advising consumers, in addition to keeping up with information on sustainability of the species he markets.

Lovewell grew up in a Cape Cod fishing family. After college in California he taught sailing on the Sea of Cortez with the National Outdoor Leadership School. The illegal fishing practices of huge international trawlers there made a big impression on him, as did the impacts the fishery had on the small coastal communities. He said,

"Local fishermen were struggling to feed their families and make a living." Lovewell went on to earn a master's in international environmental policy. He said he was drawn to fisheries as an opportunity to connect communities to the oceans. He believes the health of the environment is tied to food production and what we eat. In 2014, he won Entrepreneur of the Year in Monterey County.

RGF sells squid, Dungeness crab, Chinook salmon, sablefish, Pacific herring, abalone, oysters, white sea bass, rock cod, and other species. The most popular species are salmon and crab; least popular is herring (most people don't know what to do with them). There are five to six hundred deliveries every week, and nine hundred customers are in their system.

The purpose of community-supported fisheries is to incentivize sustainable fishing. Organizers work in collaboration with fishermen and consumers to simplify the supply chain. The consumer knows where, when, and how the fish are caught. The fishermen can get a better price, ensuring that they can earn more by fishing less. Generally RGF buys from the fishermen at a cost over the market price. RGF has fifty fishermen on their list of suppliers, with a core group of twenty-five from Fort Bragg to Monterey. RGF collects the fish, featuring a different species every week, and delivers it to a specific location where the subscribers pick it up. There are thirty-six pickup locations for consumers, from Sausalito to Salinas. RGF also gives the consumers recipes and hints on how to prepare their fish.

RGF follows the Seafood Watch guidelines for "Best Choice" or "Good Alternative." But they reserve the judgment to vary from the guidelines if they perceive that the local conditions for a species differs from the broader regional guidelines of Seafood Watch. In addition, they evaluate sustainability on a fisheries basis, which includes social and cultural health of the fishery, fishermen, and community that depend on the resource. Lovewell says, "Fishermen are out there on the ocean—they're engaging with it and they're part of it. When they bring us fish to eat, we become part of that cycle, that sphere of influence over the oceans."

There are more than forty other community-supported fisheries throughout North America; in addition, there are over fifteen community fisheries that are part of the Community Fisheries Network that is supported by the nonprofit foundation Ecotrust. Ecotrust states, "Our work in the Community Fisheries Network supports a longtime dream of local fishermen: locally owned and operated working waterfronts where they can sell directly to their community.... It also gives local people access to sustainable seafood and the opportunity to meet their fishermen."[14]

Sustainable Fishing Communities

Changes in the rules governing West Coast groundfish management took effect in 2011, allowing the assignment, sale, and purchase of catch quotas. In Monterey this meant that the quota historically caught and landed in Monterey could be sold to fishermen who would operate out of other ports. The loss of the local fishing business would be detrimental to the city's ability to maintain the infrastructure on Fisherman's Wharf. Since a working wharf enhances the draw of tourists to the adjacent downtown businesses and to Cannery Row, and provides fresh local seafood, the city government took steps to maintain its fisheries. The Monterey Fishing Community Sustainability Plan finalized in 2013 was the first to be passed in California. Community sustainability plans are cited in the major legislation regulating commercial fishing, the Magnuson-Stevens Act (MSA), as one potential method to avoid negative impacts in small fishing communities of newly instated catch share programs. A community quota fund is a legal and federally recognized entity that acquires and manages local groundfish quota for a community or region.

As part of the Monterey Fishing Community Sustainability Plan, the planners recommended developing a community quota fund, the Monterey Bay Fisheries Trust, to acquire, hold, and manage shares for the public benefit and to anchor local fisheries to the community. A similar program is developing in Morro Bay, farther down the coast.

One could question why the community has to repurchase the quota that once was a part of the common public resource. It became the fishermen's property; now the public (Monterey) has to buy it back from the fishermen. This is a tangled process.

The community-sustained fishery sounds like the system of community-based management of natural resources described by Professor Elinor Ostrom, a winner of the 2009 Nobel Prize in Economics. She disputed the notion of the "tragedy of the commons" as a misguided belief of the 1960s. She said in a magazine interview, "We have two main prescriptions: privatize the resource or make it state property with uniform rules. But sometimes the people who are living on the resource are in the best position to figure out how to manage it as a commons."[15]

Rather than privatize fishing rights, she showed that communities can sustainably manage fisheries in a common pool, given the opportunity and operating under certain conditions. These conditions include clearly defining boundaries (that of the resource and exclusion of external unentitled parties) and establishing rules regarding the harvesting of the common resource. These rules include

- Fishermen need to participate in the decision-making process.
- Monitoring of the harvest is required.
- Fishermen who violate rules have to be sanctioned.
- Where the resource's range is larger than the community, organizations are needed to plan harvest levels and establish a broader system of rules.
- There must be mechanisms to resolve conflicts.
- The community management has to be recognized by higher levels of authority.

As opposed to a tragedy of the commons, one strategy that seems to work is a resilient common-pool resource based on limited entry and self-imposed regulations, such as we've seen in the arapaima fishery of the Amazon basin (chap. 10) and in Chile's coastal shellfish (TURF; chap. 3) fishery. The Palau fishery, Sámi (chap. 5), and Coast Salish (chap. 4) fisheries are examples of indigenous community–managed common-resource fisheries that were overrun by colonization. Ironically, the successful common-pool resource fisheries, held forth as examples to justify catch share programs due to "failure of the commons," are now being promoted as "rights-based" fisheries.

Ostrom said, "I don't see the human as hopeless. There's a general tendency to presume people just act for short-term profit. But anyone who knows about small-town businesses and how people in a community relate to one another realizes that many of those decisions are not just for profit and that humans do try to organize and solve problems."[16] I would add that corporations don't always work this way.

Hauling the Net

Niaz Dorry told me in an interview, "I don't really consider myself an environmentalist. I'm more of a human-rights and economic-justice activist who sees those angles in battles traditionally coined as 'environmental battles.' I think once something becomes part of a 'green' agenda, it tends to get pigeonholed as such, and the various nuances of the issue that might actually lead to winning a battle tend to get suppressed. Take fisheries issues, for example. Most talk about it only as saving the oceans, and that turns off a big chunk of the population who can't relate to that message. I'm trying to introduce a different concept: save the small-scale fishermen, because I believe they will help us save the oceans better than the alternative—factory fishing and aquaculture."

Dorry is joined in her efforts by a growing global movement. In southern Europe, the Mediterranean Platform of Artisanal Fishers (MedArtNet) is a group of professional artisanal fishermen from different countries bordering the Mediterranean Sea (Greece, Italy, France, and Spain). They are concerned about "the future of the sea, the fisheries and the fishermen." In Chile, there are several organizations representing artisanal fishermen, including the National Council for the Defense of the Fishing Heritage of Chile (CONDEPP); in Alaska, there is the Alaska Sustainable Fisheries Trust; in Peru, the Federation for the Integration and Unification of the Artisanal Fishworkers of Peru (FIUPAP); and in South Africa, the World Forum of

Fisher Peoples. There is the global Slow Fish movement, and many more organizations. The involvement of small-scale fishermen in discussions of the future of their fisheries is leading to new and different approaches to sustain fisheries and fishermen.

There is hope for the future. We are becoming more aware of the effects of pollution and habitat destruction on fish populations. We recognize that the dumping of mining waste, industrial and farm waste, plastics, and nanoparticles, plus urban runoff of pesticides and toxic chemicals are collateral effects of our activities and technological development. We know that vast amounts of shoreline habitat are destroyed by development and shoreline armoring to mitigate rising sea levels. Indigenous peoples and fishing communities are leading the efforts to restore the damage done to fish habitat and the ocean environment. On another positive note, women increasingly play a role in fishing communities as activists and artisanal fishers.

Sometimes when I look to the sky at night, the stars remind me of the blinking lights of fishing boats on the Bay of Naples. I think back, and see them leave the harbor in Lacco Ameno, their wakes swallowed by the sea. They disappear into the blackness of the horizon where the night and sea meet. And sometimes when I look at the stars, I think of the Yámana people of Tierra del Fuego. According to their traditional beliefs, when the Yámana leave this life, they join the stars in heaven. These images put our individual lives into the context of the harmony of a greater picture. Facing the troubles we have ourselves created, there is an immediate sense of loss, but a greater view of optimism for the future. Our planet is speeding forward through the sky—we can't go backward, but we can preserve what we have that is good, and we can improve how we treat our environment. We can each contribute in our own way some small element that will help our communities solve problems, and perhaps traditions will be maintained.

Artisanal fisheries give the public an opportunity to connect with their food source, aligned with writer Michael Pollan's concept of establishing a relationship with food. Doing so engenders respect for the small planet we share, something that is needed for our own survival: take less, make responsible choices, and have a smaller impact. I am inspired by the people I met in this journey—the artisanal fishermen who have struggled to adapt to changing conditions in order to survive—and dedicated citizens who are searching for innovative solutions to complex issues as big as an ocean.

Acknowledgments

Many people read individual chapters or the whole manuscript and made constructive comments: Svanhild Andersen, Caroline Arantes, Spencer Bailey, Bjorn Barlaup, Lorenzo Ciannelli, Mike Dadswell, Alexandra Dane, Einar Eythórsson, Trevor Fay, Maria Christina Gambi, Don Gunderson, Jan Hartung, Kristin Hoelting, Ian Kirouac, Pete Knutson, Mike Macy, Mónica Orellana, Darren Porter, Fran Reichert, Art Seavey, Dan Sloan, Hal Smith, Stella Spring, Riley Starks, Fred Utter, Lee van der Voo, Knut Vollset; Chris Cokinos and the Wildbranch Writing Workshop, sponsored by *Orion Magazine*; Seth Kantner and the workshop group at the Montana Environmental Writers' Institute; and Nick O'Connell and the group at the Writers' Workshop.

Several people provided useful information or shared contacts, including Maria Christina Gambi, Vivian Montecino, Marcela Orellana, Alfredo Sfeir, Lydia Sigo, Aldo Soto, and Fred Utter.

I was fortunate to have the cooperation of the following people for interviews, boat trips, tours, and general support: Patricio Arana, Bjorn Barlaup, Patricio Bernal, Tonino Calise, Juan Carlos Cárdenas, Lorenzo Ciannelli, Paolo Ciannelli, Leesa Cobb, Mike Dadswell, Niaz Dorry, Trevor Fay, Ray Fryberg, Helge Furnes, Debra Hanuse, Maria Hines, Diego Holmgren, Rosie James, Pete Knutson, Bruce Leaman, Adam Lorio, Alan Lovewell, Matt Lumley, Svein Lyder, Wenche Lyder, Michèle Mesmain, Jo Mrozewski, Miguel Troncoso Olivares, Torulf Olsen, Troy Olson, Darren Porter, Leslie Roberson, Art Seavey, Ben Starkhouse, Jon Sundstrom, Paolo Vespoli, Knut Vollset, Shirley Wilson, and Ricardo Zarate. Mike Canino and Don Gunderson were always willing to discuss fisheries issues. Lorenzo Ciannelli, Mónica

Orellana, and Knut Vollset interpreted for me in Italy, Chile, and Norway, respectively. Thomas Johnson graphed the abalone charts. Mattias Bailey drew the figures of gear types.

Some individual chapters were published in advance by *Crosscut*, *Earth Island Journal*, *Grist*, and the *Santa Barbara Independent*. They were modified and expanded for this book.

Janet Kimball provided editorial suggestions during development of the manuscript. Johanna Rosenbohm performed the copyediting and Carrie Wicks the proofreading. I appreciate the guidance and support of Christie Henry and her team at the University of Chicago Press.

I thank my family for their patience and encouragement.

Notes

Introduction

Interview: Michèle Mesmain, July 24, 2017, by phone. Additional sources consulted on the state of Mediterranean fisheries: Colloca et al. 2004; Himes 2010; Mediterranean Platform of Artisanal Fishers 2012; Tsikliras et al. 2015; Vasilakopoulos, Maravelias, and Tserpes 2014. On indigenous fishing rights: United Nations 2008. On the state of world fisheries: Cigana 2014; United Nations Food and Agriculture Organization 2012, 2014; Greenberg 2010; Osterblom et al. 2015; Pauly 1999; van der Voo 2014; Wickens 2016. On the history of fishing: Finley 2013, 2017; Roberts 2012; Wilder 1998. On small-scale fisheries: Chuenpagdee 2011; United Nations Food and Agriculture Organization 2013, n.d.; Slow Food 2016; Tarver 2015. On catch shares: Crutchfield and Pontecorvo 1969; Food and Water Watch 2010, 2011b; Franco et al. 2014; Koslow 1982; Macinko 2014; Pinkerton and Davis 2015; *Seattle Times* staff 2013; United Nations Regional Information Centre for Western Europe 2012; van der Voo 2016.

1. I use the term *fisherman* as gender neutral and inclusive, much as *human*. Some writers have indicated that most women who fish as a way of life prefer to be called fishermen rather than fisherwomen because it doesn't set them apart from the crew and fishing community (Greenlaw 1999; Loomis 2015). Other gender-neutral terms like *fisher* or *fisher folk* are sometimes used in academic literature but are not popular on the docks.
2. Michèle Mesmain, formerly the international coordinator of Slow Fish, told me in a phone interview that in her view, some small-scale fisheries can be very damaging to the environment, whereas artisanal fishermen embrace an attitude of respect for the ecosystem and for the fish as a food source.
3. Cigana 2014.
4. Wickens 2016.
5. Wild, or even raised and released, fish caught in the ocean.
6. United Nations Food and Agriculture Organization 2012.
7. So says Dr. Ratana Chuenpagdee (2011), a researcher of fisheries and community development.

8. United Nations Food and Agriculture Organization 2013.
9. Roberts 2012.
10. Van der Voo 2014.
11. United Nations Food and Agriculture Organization 2012.
12. Van der Voo 2016.
13. United Nations Regional Information Centre for Western Europe 2012.
14. Mediterranean Platform of Artisanal Fishers 2012.
15. See "Slow Fish," http://www.slowfood.com/slowfish.

Chapter One

Interviews: Tonino Calise, Lacco Ameno, Italy, September 13, 2015; Lorenzo Ciannelli, Lacco Ameno, Italy, September 13–16, 2015; Vincenzo Caputo, Casamicciola, Italy, August 24, 2011; Paolo Ciannelli, Ischia, Italy, September 13 2015; Maria Christina Gambi, Casamicciola, Italy, August 23–24, 2011; Paolo Vespoli, Lacco Ameno, Italy, September 13–14, 2015. Additional sources consulted on pesca d'ombra: D'Anna, Badalamenti, and Riggio 1999; Pesca Sicilia, n.d.

1. Vasilakopoulos, Maravelias, and Tserpes 2014.

Chapter Two

Interviews: Bjorn Barlaup, Bergen and Bolstad Fjord, Norway, September 10, 2015; Helge Furnes, Bolstad Fjord, September 10, 2015; Knut Vollset, Bergen and Bolstad Fjord, Norway, September 10, 2015. Additional sources consulted on the status of salmon: Alexandra Morton: Gwayum'dzi, n.d.; Forseth et al. 2017; Otero et al. 2011; *Southern Pacific Review* 2015; Taranger et al. 2014; Thorstad and Forseth 2015; World Wildlife Fund 2001. On the background of salmon: Dalheim 2012; Gajic 2011; Saegrov et al. 1997. On the Vosso project: Vollset et al. 2014. On farmed versus wild salmon: Glover et al. 2017; Morton 2016; Suzuki 2014.

1. Walton 1653.
2. Disclosure: I have consulted for Uni Research as an independent program reviewer on the predation of Vosso River salmon smolts. Uni Research is a strategic research partner of the University of Bergen and is owned by the university and its research foundation.
3. World Wildlife Fund 2001.
4. Otero et al. 2011.
5. Thorstad and Forseth 2015; Bjorn Barlaup, interview with the author, Bergen and Bolstad Fjord, Norway, September 10, 2015.
6. Both Helge Furnes and Tonino Calise, reported in chap. 1, died in autumn 2015, not long after I interviewed them. In July 2017, Knut Vollset sent me a picture

of a completely wild-reared Vosso salmon that had return to the fjord system to spawn.

Chapter Three

Epigraph: Pablo Neruda, *Love: Ten Poems*, trans. Ken Krabbenhoft (New York: Miramax Books, 1995). Interviews: Patricio Arana, Chile, April 22, 2016; Patricio Bernal, Chile, May 2, 2016; Juan Carlos Cárdenas, Chile, April 20, 2016; and Miguel Troncoso Olivares, Chile, April 23, 2016. Additional sources consulted on the Ley Longueira: Nelsen 2013b; Real News Network 2013; *Southern Pacific Review* 2015; *Southern Pacific Review* 2016; On the state of Chile's fisheries: *El Economista América* 2014; Nelsen 2013a; Nelsen and Rodriguez 2013; Orensanz et al. 2006; J. Smith 2014; Subsecretaría de Pesca y Acuicultura 2016; World Fishing & Aquaculture 2013. On the Yámana and Tierra del Fuego: Cooke 1712; Furlong 1911; Gusinde 1961; Hazelwood 2001; Weddell 1825. On the economic and social conditions of Chile: Associated Press 2014; Gallardo 2016; Loken 2014; Opazo 2016. On women divers: MacEacheran 2016; Sang-Hun 2014.

1. Opazo 2016.
2. *Southern Pacific Review* 2015.
3. Julian Smith 2014.
4. Chilotes there are a mixture of the Mapuche and Chono indigenous peoples, Spaniard colonists, and French whalers. Before colonization, the island's Indians were aquatic nomads who, like the Yámana, traveled the archipelago around Chiloé in canoes, fishing and gathering shellfish. They were known to be fierce warriors.
5. Also known as the Yaghan.
6. Many readers will recognize the name Jemmy Button, as he's known in European history. His native name was O'run-del'lico. He was a Yámana Indian from Tierra del Fuego. Whether he ended up in England in 1830 of his own free will is a conundrum of history, but it is known that he traveled there after being picked up on the HMS *Beagle* under the command of Capt. Robert FitzRoy. According to various accounts, either he joined the English voluntarily or he was traded by his uncle to FitzRoy for a pearl button, thereby acquiring the surname Button. Possibly FitzRoy kidnapped him, but by the time Button figured out what was happening to him, he'd been taken far away. In England he learned British ways, met the queen, and mingled with the elites of society. As he familiarized with English clothes, Button became known as a dandy, and often preened before a mirror. However, he was anxious to return to his home and family.

After just over a year in England, he sailed with FitzRoy back to Tierra del Fuego on the epic voyage of the *Beagle*, now with Charles Darwin aboard. Darwin said of Button, "It seems yet wonderful to me, when I think over all his many good qualities, that he should have been of the same race, and doubtless partaken of the same character, with the miserable, degraded savages whom we first met here."

FitzRoy hoped that upon his return, Jemmy Button would proselytize the "savages" on the island. Instead, after a nearly three-year absence from his tribe, when Button landed in Tierra del Fuego, his fellow Yámana robbed him of most of his English accouterments. He quickly reverted to his native ways. Months later when the Englishmen encountered Jemmy Button, he was nearly naked, had lost his body fat, but he had no desire to return to England. Hazelwood 2001.

7. Gusinde 1961.
8. Windh 2016.
9. Associated Press 2014.
10. Loken 2014.
11. Loken 2014.
12. For God, glory, and gold; see Gascoigne 2000.
13. Pablo Neruda, *Selected Poems*, ed. Nathaniel Tarn (Boston: Houghton-Mifflin, 1970).
14. Sang-Hun 2014.
15. *Japan Times* 2013.

Chapter Four

Interviews: Ray Fryberg, Tulalip Reservation, December 3, 2014, March 31, 2015; Debra Hanuse, interview by the author, Alert Bay, British Columbia, April 1, 2015; Rosie Cayou James, Anacortes, Washington, January 26, 2015; Adam Lorio, Anacortes, Washington, January 26, 2015; Jo Mrozewski, Alert Bay, British Columbia, March 31, 2015; Lydia Sigo, Suquamish Museum, Port Madison Indian Reservation, Washington, February 17, 2015. Additional sources consulted on the salmon ceremony: Amoss 1987; Archibald 2008; Dover 2013; Gunther 1928; Janes 2003. On Salish history: W. Angelbeck 2009; B. Angelbeck and McLay 2011; Dover 2013; Elmendorf 1993; Erlandson, Moss, and Lauriers 2008; Haeberlin and Gunther 1930; Kirk 1986; McDaniel 2004; Moss 2011; Stein 2000; Suttles 1954; Swan (1857) 1972; Thrush, n.d.; *Washington Historical Quarterly* 1934; and a visit by the author to Hibulb Cultural Center and Natural History Preserve, Tulalip, Washington, February 29, 2015. On the history of the treaty and fishing: Northwest Treaty Tribes 2013; Pembroke 1979; Schwartz 2015b; Stewart 1977; Wilkinson 2000. On the natural history of salmon and the Pacific Northwest: Barron and Anderson 2011; Hebda and Matthews 1984; Lichatowich 1999, 2013; Montgomery 2003. On Fort Langley: Barman 1999;

Klan 1999; MacLachlan 1998; McKelvie 1947. On the Boldt decision: Knutson 1989; Tizon 1999; Woods 2005.

1. From the Program of the Tulalip Tribes Salmon Ceremony, June 13, 2015.
2. Wilkinson 2000.
3. "Two-eyed seeing," or etuaptmumk, is the gift of multiple perspectives, according to Mi'kmaq elder Albert Marshall (Institute for Integrative Science and Health 2012). This is not a Coast Salish term, but in my opinion is a useful expression for the idea of co-learning multiple cultural perspectives.
4. James Murray Yale was a stout man, cantankerous, and with a short temper that matched his equally short stature. The Indians called him "Little Yale." Yale drove a hard bargain with the Indians and generally swapped about a half-penny's worth of goods for each fish. The Snohomish and other Coast Salish bands felt bitterness toward Yale because he had conducted a raid on the Klallam tribe a few years before and killed twenty-two of the Indians.

    Yale was unlucky in love. His first wife left him for another man, and his second, T'seeyiya, was the daughter of a powerful Cowichan elder, but as it turned out, she had been already married off by her father to another man.

    In his twenty-six years as the chief trader at Fort Langley, James Murray Yale supervised a growing demand for salted salmon that was delivered to Hawaii and the East Coast. He also procured otter and beaver pelts for the Hudson's Bay Company, constantly demanding more furs from the Indians.
5. The story of Qaba'xad is not a traditional Native American story. He was a historical character mentioned in Elmendorf 1993. I searched extensively for his ancestors with the assistance of the Tulalip Museum staff, but was unable to find records of living ancestors. My story of Qaba'xad is, of course, a creative narrative based on historical and cultural backgrounds. I have told it from the perspective of James Murray Yale, as he might have seen it.
6. The story of the Yuyubaĉ was told to me by Ray Fryberg of the Snohomish tribe.
7. The Snohomish people used nets to beach-seine and to gillnet, and in fishing weirs to catch salmon. Some elders reported the use of reefnets or reefnet-like capture methods in camps on Lopez Island. However, the Northern Straits Salish—mainly the Lummi and Samish of the San Juan Islands—were the only well-documented users of reefnets. It could be the Snohomish members married into Northern Straits families and fished with the other tribes, although they spoke different dialects. One might speculate this was more common postcontact, when the Salish populations were depleted by epidemics. Processing large amounts of salmon is labor-intensive and requires a critical mass of people.
8. From a display at the Hibulb Cultural Center and Natural History Preserve, Tulalip Reservation, February 29, 2015.
9. The First Salmon ceremony was probably different for each tribe and maybe for

every village within a tribe. Ray Fryberg, executive director of the Natural Resources Department of the Tulalip Reservation, shared his view in an interview with me that the details of the ritual itself weren't as important as the meaning behind it.

10. Written descriptions of the ceremony of the Snohomish were made in 1924 and published in 1928 by Erna Gunther. By this time the ceremony had largely been lost, and probably was stifled in the mid-1860s, when missionaries found out about the tribe's rituals and banned them. Many of the rituals were kept secret and not revealed to non-Indians in order to protect them. The northern tribes in British Columbia had more elaborate rituals still, and when Kirk had examined them, it seems these tribes maintained more of their traditions. For example, the old people and children were bathed, painted with red ochre, and dusted with down feathers. They were more isolated and less affected by the settlers. Presumably the early postcontact ceremonies of the Snohomish were more elaborate than those described by Kirk. The story of the First Salmon ceremony was also shared with me by Ray Fryberg. It is told beautifully by Harriette Shelton Dover in her book *Tulalip, from My Heart*. I suggest that anyone retelling the narrative should consult with Dover's telling of the story.

11. Other tribes, such as the Lummi, made their nets from willow saplings and dried nettles.

12. Stewart 1977.

13. Stevens's ultimate goal was to move all Native Americans in the territory onto one reservation, but he was killed in battle by a gunshot wound during the Civil War before he could accomplish his goal.

14. The amount of land ceded by the tribes in the Point Elliott Treaty isn't readily available. I estimate the amount at more than four million acres from a Washington State map showing the ceded property: "Tribal Ceded Areas in Washington State: Washington Department of Fish and Wildlife Interpretation," http://wdfw.wa.gov/hunting/tribal/wa_tribal_ceeded_lands.pdf, accessed July 13, 2017. Tribes signing the treaty received 53,168 acres for their reservations. The Tulalip tribe received 22,459 acres.

15. Quoted in Schlosser 1978.

16. See Wilkinson 2000.

17. Thrush, n.d.

18. Kirk 1986.

19. Elmendorf 1993.

20. The differing versions of this historical event are of note. It was fairly unusual for the Coast Salish to unite hundreds or thousands of warriors, so it's probable that they describe the same battle. One document (*Washington Historical Quarterly* 1934) says that Kitsap led a coalition against the Cowichans in 1825 and they lost. Elmendorf (1993) describes the Battle of Maple Bay as occurring

in about 1845 between an alliance of Coast Salish warriors and a raiding party of Kwakiutl. McKelvie (1947) describes a great battle outside of Fort Langley's walls (there is a Maple Ridge nearby) in 1837 that broke the power of the Yuculta (southern Kwakiutl). They had come to destroy the Whaneetum (whites). Thousands died. In this version, guns and cannons were prevalent. The Kwantlen (Salish of the lower Fraser River) pursued the attackers into the forest and finished them off.

21. Harriette Shelton Dover (2013), an elder of the Tulalip tribes, told the story of her people in the book *Tulalip, from My Heart*. Her story of the people weaves together with the natural histories of Pacific salmon and cedar. Other Indian people in the Pacific Northwest had similar linkages, which prompted the development of a rich cultural heritage. With the arrival of Europeans, the braid of people, salmon, and cedar came unwoven. The traditional Coast Salish salmon fishery changed. Dover herself was a key personality in reviving the First Salmon ceremony as a cultural tradition of the Tulalips to remind them of their heritage as a fishing people.
22. Quoted in Wilkinson 2000.
23. See http://www.namgis.bc.ca.

Chapter Five

Interviews: Svein Lyder, Veidnes, Norway, September 7, 2015; Wenche Lyder, Kvalsund, September 6–7, 2015; Torulf Olsen, Kvalsund, Norway, September 6–9, 2015. Additional sources consulted on the region's prehistory: Broadbent 2010; Fitzhugh 1975; Gjessing 1975; Kuoljok and Utsi 1993; MacAulay 1998; Meriot 1984; Odner 1992. On Sámi history: Bergman et al. 2008; Evjen and Hansen 2009; Hansen and Olsen 2014; Westerdahl 2010. On Sámi fishing history: Boekraad 2013; Eythórsson 2003; Pedersen 2012. On contemporary Norwegian fisheries: Bailey 2013; Havfisk 2013; Henriksen 2014; Myklebust 2014; Norwegian Environmental Agency 2014; Odendahl 2016; Undercurrent News 2014. On king crab: Barentsinfo.org, n.d. On recent Sámi fishing rights: Brattland 2010; Broderstad and Eythórsson 2014; Eythórsson and Brattland 2012; Lätsch 2012; C. Smith 2014; United Nations Economic and Social Council 2010. On changing fisheries: Bojer 1923. On the Bering Sea dory fishery for cod: Follansbee 2006; Shields 2001.

1. Several works claim that the Sámi, rather than descending directly from the Stone Age people, invaded northern Norway from the east after the major population of Scandinavians had already arrived. Others claim that the Lapps (herders) were small dwarfish people and the Finns were invading hunter-gatherers.
2. *Bivdi* is a Sámi word meaning "harvesting" or "hunting."

3. United Nations 2008.
4. *Sjøsamene* in Norwegian.
5. MacAulay 1998.
6. Anthropologists classify the Sámi groups differently, according to language and culture instead.
7. Some historians believe that reindeer herding began in the sixteenth century, while others say its origins are much earlier, AD 200 to 1000 (Bergman et al. 2008).
8. I am half Swedish. Through genetic tests I've learned that my ancestry has distant traces of Sámi and American Indian, and strong evidence of shared genes with northern European hunter-gatherers and Finns.
9. Finn was the original Norwegian name for the Sámi. Later, the Sámi were referred to as Finns for the coastal hunter-gatherers and Lapps for the reindeer herders (Hansen 2011).
10. Skogvang 2014.
11. Naturvernforbundet 2015.
12. Carson 1962.
13. Bailey 2013.
14. A sense of the value of the fish quotas can be seen in the recent sale of one of Havfisk's boats for thirty-three million Norwegian kroner. The value of the boat was one million Norwegian kroner; the rest was the value of the quota.
15. It could be that Røkke will give some back. In 2017 he announced that he will build a multimillion-dollar vessel for marine research.
16. Eythórsson 2003.
17. Bojer 1923.
18. The Alaskan cod were salted to preserve them and not dried like the Norwegian cod due to the wet conditions along the shoreline of the North Pacific Ocean.
19. Follansbee 2006; Cobb 1927; Shields 2001.

Chapter Six

Interviews: Ian Kirouac, Lummi Island, Washington, July 28, 2015; John Sundstrom, by telephone, August 7, 2015. Additional sources consulted on the Lummi Island Wild Company background: Lummi Island Wild, n.d.; Scheer 2007. On reefnetting history: Boxberger 2000; S. K. Campbell and Butler 2010; Clark 1980; Easton 1990; James 2013. On the modern Lummi fishery: Mapes 2016; Northwest Treaty Tribes 2014; Schwartz 2015b. On the coal terminal: Schwartz 2015a. On sea gypsies and fish listeners: V. Brown 2013; Ivanoff 2005.

1. Easton 1990.
2. There is one reefnet gear set in the San Juan Islands reportedly owned by Indian fishermen.

3. Schwartz 2015b.
4. Schwartz 2015a.
5. Lummi Island Wild, n.d.
6. Survival International, n.d.
7. Ivanoff 2005.
8. *Star Online* 2014.
9. V. Brown 2013.

Chapter Seven

Interviews: Mike Dadswell, Bramber, Nova Scotia, Canada, June 10, 2016; Matt Lumley, Halifax, Nova Scotia, Canada, June 11, 2016; Darren Porter, Bramber, Nova Scotia, Canada, June 10, 2016. Additional sources consulted on power company project developments and status: F. Campbell 2016; *Chronicle Herald* 2016; Daborn 2016a; Dempsey 2016; Elliott 2014; Riley 2016; *Shoreline Journal* 2016; Wright 2016. On FORCE and tidal power companies: Cape Sharp Tidal, n.d.; Ecology Action Centre 2016; Fundy Ocean Research Center for Energy 2016, n.d.; McLean 2016. On the environmental impacts of turbines: Daborn 2016b; Wilson et al. 2006. On experiences in a West African fishery: Roberson 2015.

1. Dempsey 2016.
2. Cape Sharp Tidal, n.d.
3. Fundy Ocean Research Center for Energy 2016.
4. Ecology Action Centre 2016; McLean 2016 (letter from DFO).
5. *Shoreline Journal* 2016.
6. Darren Porter, personal communication, reporting on a meeting with Margaret Miller on August 23, 2016.
7. Ecology Action Centre 2016.
8. *Chronicle Herald* 2016.
9. Riley 2016.
10. Elliott 2014.
11. Daborn 2016a, 2016b.
12. Wright 2016.
13. The first of the Cape Sharp Tidal turbines went in the water on November 8, 2016. It was removed in June 2017, as it was malfunctioning. As of this writing, it has not been redeployed yet.
14. Roberson 2015.
15. Roberson 2015.
16. Roberson 2015.

Chapter Eight

Interviews: Maria Hines, by telephone, October 30, 2013; Peter Knutson, Seattle, Washington, June 7, 2013, October 31, 2014, November 4, 2014, October 25, 2015, and February 6, 2016; Bruce Leaman, interview by the author, Seattle, Washington, August 6, 2015. Additional sources consulted on salmon conflicts and politics: Knutson 1989, 2017; Murakami 1995. On the Loki Fish Company: Loki Fish Company, n.d.; Nickel-Kailing 2015. On the conundrum of the Pacific halibut: Brooks 2015.

1. For example, see "Sustainable Fishing and Environmental Stewardship from Washington to Alaska," Vimeo video, 7:20, posted by Pangeality Productions, 2010, http://vimeo.com/11448118.
2. Borderías and Sánchez-Alonso 2011.
3. Murakami 1995.
4. See http://www.mariahinesrestaurants.com.

Chapter Nine

Interviews: Trevor Fay, Monterey, California, January 5, 2014, June 10, 2014, April 15, 2016; Dorothy Garfield, by telephone, August 9, 2016; Art Seavey, Monterey, California, January 5, 2014. Additional sources consulted on farming abalone: Abraham 2009; Hale 2009; Muhlke 2009; St. Fleur 2014; Thomas 2014; Wu 2007. On the kelp issue: King and DeVogelaere 2000; Monterey Bay National Marine Sanctuary 2014; Spicuzza 1998. On the 'Namgis tradition of ocean farming: Groesbeck 2014; Moss 2011. On Cuttyhunk Shellfish Farms: Cuttyhunk Shellfish Farms, n.d.

1. Coincidentally I wrote a book about the *Western Flyer* (*The* Western Flyer*: Steinbeck's Boat, the Sea of Cortez, and the Saga of Pacific Fisheries*, Chicago: University of Chicago Press, 2015).
2. Now named American Abalone Farms.
3. St. Fleur 2014.

Chapter Ten

Interview: Caroline Arantes, by telephone, March 2, 2016. Additional sources consulted on the natural history and exploitation of arapaima: Fernandes 2006; Miranda-Chumacero et al. 2012. On the Amazon fishery: Campos-Silva and Peres 2016; Cavole, Arantes, and Castillo 2015; McGrath et al. 2008. On farming arapaima: Amazone Company, n.d.; Chauvin 2012; Seafood News 2015; Wong 2014; Zambito 2014; and pers. comm., Aldo Soto to the author, email, January

15, 2014. On cooperative fishing with dolphins: Daura-Jorge et al. 2012; Roman 2013; Zappes et al. 2011. On Veta la Palma: Veta la Palma, n.d.

1. The genus Arapaima is undergoing taxonomic revision. *Arapaima gigas* may include several species.
2. Here is a nice video of the fishing method: "Projeto Várzea: Manejo do Pirarucu no Baixo Amazonas," YouTube video, 15:24, posted by WWF-Brasil, September 3, 2012, https://www.youtube.com/watch?v=4T7HBlhdgJM.
3. "Jaguar Fishing Amazon Giant Pirarucu (Arapaima-gigas)," Vimeo video, 2:07, posted by A. D'Silva, 2011, http://vimeo.com/23655267.
4. "Projeto Várzea: Manejo do Pirarucu no Baixo Amazonas."
5. McGrath et al. 2008.
6. There is an argument about the cost of feeding fish to other fish. Why use three kilograms of anchovy to make one kilogram of paiche? The answer to that is most people won't eat anchovies. We mostly use anchovies to grow chicken or salmon. Fish meal from anchovies is also used to fertilize vegetables. Whole Foods aims for a 1:1 ratio of feed fish to cultured fish in the products they sell.
7. Zarate has since left the restaurants but is still cooking in Southern California.
8. According to Amazone company reports, there are high levels of omega-3s in arapaima, presumably coming from the portion of its feed that is anchovy. High amounts of omega-3s are generally not linked to warm-freshwater species.
9. Zambito 2014.

Chapter Eleven

Interviews: Leesa Cobb, Port Orford, Oregon, September 4, 2014; Niaz Dorry, Seattle, Washington, July 26, 2014; Donald Gunderson, Seattle, Washington, April 4, 2017; Alan Lovewell, Monterey, California, May 18, 2015; Michèle Mesmain, by telephone, July 24, 2017. Additional sources consulted on Palau: Johannes 1981. On catch shares: Food and Water Watch 2011a; Hill 2017; Lawrence 2016a, 2016b; McDonald 2013; National Oceanic and Atmospheric Administration 2010; Raines 2016; Stoll 2016; van der Voo 2016. On community-based quota: P. L. Brown 2014; Ostrom 1990. On community-supported fisheries: Dorry 2015; Port Orford Sustainable Seafood, n.d.; Real Good Fish, n.d. On science and certification: Fairbrother 2012; Fundaro 2016; Opitz et al. 2016.

1. Named by *Time* in 1998 as a "Hero of the Planet," Dorry hasn't always been treated politely by the fishing community. She told me about a confrontation with an Alaskan fisherman at a North Pacific Fishery Management Council meeting when she worked for Greenpeace. He threatened her with a closed fist, shaking it close to her face. She said to herself, "This time I'm actually going to get hit." The mantra she'd learned in Greenpeace's training played over and over in her mind: "You are a tree . . . even if he hits you, he can't knock you down."

Then she saw a big ring on his hand and thought, "This could hurt ... but you are a tree ..." Dorry stood strong in her belief, and the fisherman eventually backed down.

2. McDonald 2013.
3. McDonald 2013.
4. Lawrence 2016a. In March 2017 Rafael pleaded guilty to the charges of evading quotas by falsifying records; smuggling profits to Portugal; and income tax evasion. Hill 2017.
5. Lawrence 2016b.
6. North Pacific Fishery Management Council 2016.
7. *Seattle Times* staff 2013.
8. Bernton 2014.
9. Fundaro 2016.
10. Raines 2016; van der Voo 2016.
11. Mendenhall 2015; van der Voo 2016.
12. North Pacific Fishery Management Council 2016.
13. Pinkerton, Olsen, et al. 2016.
14. Ecotrust, n.d.
15. Korten 2010.
16. Korten 2010.

References

Abraham, Kera. 2009. "Entrepreneurs and Scientists Team Up to Produce a Prettier, Healthier Abalone." *Monterey County Weekly*, February 5.
Alexandra Morton: Gwayum'dzi. N.d. http://www.alexandramorton.ca. Last accessed July 13, 2017.
Amazone Company. N.d. http://www.amazone.com.pe. Last accessed July 13, 2017.
Amoss, Pamela T. 1987. "The Fish God Gave Us: The First Salmon Ceremony Revived." *Arctic Anthropology* 24 (1): 56–66.
Angelbeck, B., and E. McLay. 2011. "The Battle at Maple Bay: The Dynamics of Coast Salish Political Organization through Oral Histories." *Ethnohistory* 58 (3): 359–92. https://doi.org/10.1215/00141801-1263821.
Angelbeck, William O. 2009. "Power, Practice, Anarchism and Warfare in the Coast Salish Past." PhD diss., Department of Anthropology, University of British Columbia.
Archibald, Jo-ann. 2008. *Indigenous Storywork: Educating the Heart, Mind, Body, and Spirit*. Vancouver: University of British Columbia Press.
Associated Press. 2014. "OECD Says Chile Has Widest Inequality Gap." CBS Money Watch, March 18. http://www.cbsnews.com/news/oecd-says-chile-has-widest-inequality-gap/.
Bailey, Kevin. 2013. *Billion-Dollar Fish: The Untold Story of Alaskan Pollock*. Chicago: University of Chicago Press.
Barentsinfo.org. N.d. "Red King Crab." http://www.barentsinfo.org/Contents/Nature/Animals-and-plants/King-crab. Last accessed December 4, 2016.
Barman, Jean. 1999. "Family Life at Fort Langley." *British Columbia Historical News* 32 (4): 16–30.
Barron, John A., and Lesleigh Anderson. 2011. "Enhanced Late Holocene ENSO/PDO Expression along the Margins of the Eastern North Pacific." *Quaternary International* 235 (1–2): 3–12. https://doi.org/10.1016/j.quaint.2010.02.026.

Bergman, Injela, Lars Liedgren, Lars Ostlund, and Ollie Zackrisson. 2008. "Sami Residence Patterns in Fennoscandia Alpine Areas around AD 1000." *Arctic Anthropology* 45 (1): 97–110.

Bernton, Hal. 2014. "American Seafoods to Pay 1.75M Penalty for Weighing Inaccuracies." *Seattle Times*, October 15. http://www.seattletimes.com/business/american-seafoods-to-pay-175m-penalty-for-weighing-inaccuracies/.

Boekraad, Gerarda Maria Doeke. 2013. "Ecological Sustainability in Traditional Sami Belief and Rituals." Master's thesis, Department of Archaeology, History, Cultural Studies and Religion, University of Bergen.

Bojer, Johan. 1923. *The Last of the Vikings*. Translated by Jessie Muir. New York: Century.

Borderías, Antonio J., and Isabel Sánchez-Alonso. 2011. "First Processing Steps and the Quality of Wild and Farmed Fish." *Journal of Food Science* 76 (1): R1–R5. http://www.ncbi.nlm.nih.gov/pmc/articles/PMC3038323.

Boxberger, Daniel. 2000. *To Fish in Common: The Ethnohistory of Lummi Indian Salmon Fishing*. Seattle: University of Washington Press.

Brattland, Camilla. 2010. "Mapping Rights in Coastal Sami Seascapes." *Arctic Review on Law and Politics* 1 (1): 28–53.

Broadbent, Noel D. 2010. *Lapps and Labyrinths*. Washington, DC: Smithsonian Institute Scholarly Press.

Broderstad, Else Grete, and Einar Eythórsson. 2014. "Resilient Communities? Collapse and Recovery of a Social-Ecological System in Arctic Norway." *Ecology and Society* 19 (3). https://doi.org/10.5751/es-06533-190301.

Brooks, James. 2015. "Bering Sea Halibut Bycatch Cut Leaves Both Trawlers and Halibut Fishermen Unhappy." *Juneau Empire*, June 8. http://juneauempire.com/state/2015-06-08/bering-sea-halibut-bycatch-cut-leaves-both-trawlers-and-halibut-fishermen-unhappy.

Brown, Patricia Leigh. 2014. "Creating a Safe Harbor for a Village Heritage." *New York Times*, July 6. http://www.nytimes.com/2014/07/07/us/creating-a-safe-harbor-for-a-village-heritage.html.

Brown, Victoria. 2013. "The Last Fish Listeners." *Star Online*, September 20. http://www.thestar.com.my/Opinion/Online-Exclusive/Behind-The-Cage/Profile/Articles/2013/09/20/The-Fish-listeners/.

Campbell, Francis. 2016. "Mi'kmaq Concerned about Fundy Tidal Power Project." *Local Express*, July 5. https://www.localxpress.ca/local-news/mikmaq-concerned-about-fundy-tidal-power-project-330770.

Campbell, S. K., and V. L. Butler. 2010. "Archaeological Evidence for Resilience of Pacific Northwest Salmon Populations and the Socioecological System over the Last ~7,500 Years." *Ecology and Society* 15 (1): 17.

Campos-Silva, João Vitor, and Carlos A. Peres. 2016. "Community-Based Management Induces Rapid Recovery of a High-Value Tropical Freshwater Fishery." *Scientific Reports* 6:34745. https://doi.org/10.1038/srep34745.

Cape Sharp Tidal. N.d. "The Cape Sharp Tidal Project." http://capesharptidal.com/about-the-project. Last accessed June 18, 2016.

Carson, Rachel. 1962. *Silent Spring*. Boston MA: Houghton Mifflin.

Cavole, L. M., C. C. Arantes, and L. Castillo. 2015. "How Illegal Are Tropical-Small Scale Fisheries? An Estimate for Arapaima in the Amazon." *Fisheries Research* 168:1–5.

Chauvin, Lucien. 2012. "Foodie Alert: Would You Have a Living Fossil for Dinner?" *Time*, August 10. http://world.time.com/2012/08/10/foodie-alert-would-you-have-a-living-fossil-for-dinner/.

Chronicle Herald. 2016. "Cape Sharp Tidal Bay of Fundy Turbines Near Ready for Testing, Fishermen Wary." May 3. http://thechronicleherald.ca/nova scotia/1361792-cape-sharp-tidal-bay-of-fundy-turbines-near-ready-for-testing-fishermen-wary.

Chuenpagdee, Ratana, ed. 2011. *World Small-Scale Fisheries*. Delft, The Netherlands: Eburon Academic Publishers.

Cigana, Caio. 2014. "Peixes importados fisgam consumidores" [Imported fish squeezes consumer]. *Diario Catarinense*, April 5. http://zerohora.clicrbs.com.br/rs/economia/noticia/2014/04/peixes-importados-fisgam-consumidores-4466639.html.

Clark, Richard. 1980. *Point Roberts, USA: The History of a Canadian Enclave*. Bellingham, WA: Textype.

Cobb, John N. 1927. "The Pacific Cod Fisheries." Bureau of Fisheries Document No. 1014, 385–99. Washington, DC: US Government Printing Office.

Colloca, F., V. Crespi, S. Cerasi, and S. R. Coppola. 2004. "Structure and Evolution of the Artisanal Fishery in a Southern Italian Coastal Area." *Fisheries Research* 69:359–69.

Cooke, Edward. 1712. *A Voyage to the South Sea and around the World*. London: B. Lintot and R. Gosling.

Crutchfield, J. A., and G. Pontecorvo. 1969. *The Pacific Salmon Fisheries: A Study of Irrational Conservation*. Baltimore, MD: Johns Hopkins Press.

Cuttyhunk Shellfish Farms. N.d. http://cuttyhunkshellfish.com/. Last accessed July 13, 2017.

Daborn, Graham. 2016a. "Tidal Energy: Where Are We Going?" *Athenaeum*, March 3. http://theath.ca/opinions/tidal-energy-where-are-we-going/.

———. 2016b. "Tidal Power from Fundy—Separating Fact from Fiction." *Chronicle Herald*, October 14. http://thechronicleherald.ca/opinion/1405 990-opinion-tidal-power-from-fundy-%E2%80%94-separating-fact-from-fiction.

Dalheim, Line. 2012. "Into the Wild and Back Again: Hatching 'Wild Salmon' in Western Norway." Master's thesis, Department of Social Anthropology, University of Oslo.

D'Anna, G., F. Badalamenti, and S. Riggio. 1999. "Traditional and Experimental Floating Fish Aggregating Devices in the Gulf of Castellammare (NW Sicily): Results from Catches and Visual Observations." *Scientia Marina* 63 (3–4): 209–18.

Daura-Jorge, F. G., M. Cantor, S. N. Ingram, D. Lusseau, and P. C. Simoes-Lopes. 2012. "The Structure of a Bottlenose Dolphin Society Is Coupled to a Unique Foraging Cooperation with Artisanal Fishermen." *Biology Letters* 8:702–5.

Dempsey, Stephen. 2016. "Tidal-Power Development Is Already a Breakthrough." *Chronicle Herald*, May 30. http://thechronicleherald.ca/opinion/1368437-opinion-tidal-power-development-is-already-a-breakthrough.

Dorry, Niaz. 2015. "'Niaz' Top 10 Myths About Seafood, Fisheries, and Marine Conservation." NAMA (Northwest Atlantic Marine Alliance), August 20. http://whofishesmatters.blogspot.com/2015/08/niaz-top-ten-myths-about-fisheries.html.

Dover, Harriette Shelton. 2013. *Tulalip, from My Heart: An Autobiographical Account of a Reservation Community*. Seattle: University of Washington Press.

Duncan, David James. 1995. *River Teeth: Stories and Writings*. New York: Random House.

Easton, N. Alexander. 1990. "The Archaeology of Straits Salish Reef Netting: Past and Future Research Strategies." *Northwest Anthropological Research Notes* 24 (2): 161–77.

Ecology Action Centre. 2016. "Initial Statement Regarding the Cape Sharp Tidal Demonstration Project in the Bay of Fundy." May 24. https://www.ecologyaction.ca/files/images-documents/Ecology%20Action%20Centre%20Statement%20Regarding%20the%20Cape%20Sharp%20Tidal%20Demonstration%20Project%20in%20the%20Bay%20of%20Fundy%20(2).pdf.

Ecotrust. N.d. "Next Generation Fishing Communities." https://ecotrust.org/project/community-fisheries/. Last accessed July 24, 2017.

El Economista América. 2014. "48% of Chile's Fisheries Are Overexploited or Depleted." January 4. http://www.eleconomistaamerica.cl/empresas-eAm-chile/noticias/5671199/04/14/El-48-de-las-pesquerias-en-Chile-estan-sobreexplotadas-o-agotadas-.html.

Elliott, Wendy. 2014. "Acadia Prof Questions Reality of Fish-Friendly Turbines." *Nova News*, October 11. http://www.novanewsnow.com/News/Local/2014-10-11/article-3900852/Acadia-prof-questions-reality-of-fish-friendly-turbines/1.

Elmendorf, W. W. 1993. *Twana Narratives: Native Historical Accounts of a Coastal Salish Culture*. Seattle: University of Washington Press.

Erlandson, J. M., M. L. Moss, and M. D. Lauriers. 2008. "Life on the Edge: Early Maritime Cultures of the Pacific Coast of North America." *Quaternary Science Reviews* 27:2232–45.

Evjen, Bjorg, and Larsivar Hansen. 2009. "One People—Many Names: On Different Designations for the Sami Population in the Norwegian County of Nordland through the Centuries." *Continuity and Change* 24 (2): 211–43.

Eythórsson, Einar. 2003. "The Coastal Sami: A 'Pariah Caste' of the Norwegian Fisheries?" In *Indigenous Peoples: Resource Management and Global Rights*, edited by Svein Jentoft, Henry Minde, and Ragnar Nilsen, 149–62. Delft, The Netherlands: Eburon Academic Publishers.

Eythórsson, Einar, and Camilla Brattland. 2012. "New Challenges to Research on Local Ecological Knowledge: Cross-Disciplinarity and Partnership." In *Fishing People of the North*, edited by C. Carothers et al., 131–52. Fairbanks: Alaska Sea Grant, University of Alaska.

Fairbrother, Alison. 2012. "Science for Hire: Why Industry's Deep Pockets May Be Depleting the Last of Our Fisheries." AlterNet, November 17. http://www.alternet.org/environment/science-hire-why-industrys-deep-pockets-may-be-depleting-last-our-fisheries.

Fernandes, D. 2006. "'More Eyes Watching . . .': Community-Based Management of the Arapaima (*Arapaima gigas*) in Central Guyana." In *Eleventh Biennial Conference of the International Association for the Study of Common Property (IASCP)*. June 19–23, Bali, Indonesia. http://dlc.dlib.indiana.edu/dlc/bitstream/handle/10535/711/Fernandes_Damian.pdf.

Finley, Carmel. 2013. *All the Fish in the Sea: Maximum Sustainable Yield and the Failure of Fisheries Management*. Chicago: University of Chicago Press.

———. 2017. *All the Boats on the Ocean: How Government Subsidies Led to Global Overfishing*. Chicago: University of Chicago Press.

Fitzhugh, W, ed. 1975. *Prehistoric Maritime Adaptations of the Circumpolar Zone*. The Hague: Mouton Press.

Follansbee, Joe. 2006. *Shipbuilders, Sea Captains, and Fishermen: The Story of the Schooner Wawona*. New York: iUniverse.

Food and Water Watch. 2010. "Catch-and-Trade Catastrophes: Failures in Fishery Quota Programs." September. https://www.foodandwaterwatch.org/sites/default/files/catch_trade_catastrophes_ib_sept_2010.pdf.

———. 2011a. "A Closer Look at Catch Shares in the United States: The Gulf of Mexico." November. https://www.foodandwaterwatch.org/sites/default/files/gulfcatchshares.pdf.

———. 2011b. "Fish, Inc.: The Privatization of U.S. Fisheries through Catch

Share Programs." June. https://www.foodandwaterwatch.org/sites/default/files/fish_inc_report_june_2011.pdf.

Forseth, Torbjørn, Bjørn T. Barlaup, Bengt Finstad, Peder Fiske, Harald Gjøsæter, Morten Falkegård, Atle Hindar, Tor Atle Mo, Audun H. Rikardsen, Eva B. Thorstad, Leif Asbjørn Vøllestad, and Vidar Wennevik. 2017. "The Major Threats to Atlantic Salmon in Norway." *ICES Journal of Marine Science* 74 (6): 1496–1513. https://doi.org/10.1093/icesjms/fsx020.

Franco, Jennifer, Pietje Vervest, Timothe Feodoroff, Carsten Pedersen, Ricarda Reuter, and Mads Christian Barbesgaard. 2014. "The Global Ocean Grab: A Primer." Transnational Institute, September 2. https://www.tni.org/en/publication/the-global-ocean-grab-a-primer.

Fundaro, Steve. 2016. "Pelagic Red Crabs Return to Monterey, Bad Sign for Fishermen." *KION News*, May 24. http://www.kionrightnow.com/news/local-news/pelagic-red-crabs-return-to-monterey-bad-sign-for-fishermen/39689288.

Fundy Ocean Research Center for Energy. 2016. "Environmental Effects Monitoring Programs." March. http://fundyforce.ca/wp-content/uploads/2012/05/FORCE-EEMP-2016.pdf.

———. N.d. http://fundyforce.ca/. Last accessed July 30, 2017.

Furlong, Charles W. 1911. "Cruising with the Yahgans." *Outing Magazine* 58: 3–17.

Gajic, Nevena. 2011. "Human Dimensions of Natural Resource Management for the Vosso Wild Salmon Population." Master's thesis, Department of Industrial Economics and Technology Management, Norwegian University of Science and Technology.

Gallardo, Alejandro Martinez. 2016. "Opinión: La paradoja de Chile, el país más depresivo del mundo" [The paradox of Chile, the most depressed country in the world]. *El Mostrador*, April 22. http://www.elmostrador.cl/noticias/pais/2016/04/22/opinion-la-paradoja-de-chile-el-pais-mas-depresivo-del-mundo/.

Gascoigne, John. 2000. "Motives for European Exploration of the Pacific in the Age of Enlightenment." *Pacific Science* 54 (3): 222–37.

Gjessing, G. 1975. "Maritime Adaptations in Northern Norway's Prehistory." In *Prehistoric Maritime Adaptations of the Circumpolar Zone*, edited by W. Fitzhugh, 87–100. The Hague: Mouton.

Glover, K. A., M. F. Solberg, P. McGinnity, K. Hindar, E. Verspoor, M. W. Coulson, M. M. Hansen, H. Araki, O. Skaala, and T. Svasand. 2017. "Half a Century of Genetic Interaction between Farmed and Wild Atlantic Salmon: Status of Knowledge and Unanswered Questions." *Fish and Fisheries*, March 10, 1–38. https://doi.org/10.1111/faf.12214.

Greenberg, Paul. 2010. *Four Fish: The Future of the Last Wild Food.* New York: Penguin.

Greenlaw, Linda. 1999. *The Hungry Ocean: A Swordboat Captain's Journey.* New York: Hachette.

Groesbeck, A. S., K. Rowell, D. Lepofsky, and A. K. Solomon. 2014. "Ancient Clam Gardens Increase Shellfish Production: Adaptive Strategies from the Past Can Inform Food Security Today." *PLOS ONE* 91235. https://doi.org/10.1371/journal.pone.0091235.

Gunther, Erna. 1928. "A Further Analysis of the First Salmon Ceremony." *University of Washington Publications in Anthropology* 2 (5): 129–73.

Gusinde, Martin. 1961. *The Yamana: The Life and Thought of the Water Nomads of Cape Horn.* Translated by Frieda Schütze. New Haven, CT: Human Relations Area Files.

Haeberlin, Hermann, and Erma Gunther. 1930. "Indians of Puget Sound." *University of Washington Publications in Anthropology* 4 (1): 1–84.

Hale, Mike. 2009. "Science Meets Seafood, and the Result Is Delicious." *Monterey Herald*, January 22.

Hansen, Lars Ivar, and Bjørnar Olsen. 2014. *Hunters in Transition: An Outline of Early Sámi History.* Boston: Brill.

Havfisk. 2013. "Havfisk ASA Sells Vessel and Quota." Globe Newswire, July 3. http://inpublic-test.globenewswire.com/2013/07/03/HAVFISK+ASA+sells+vessel+and+quota+HUG1714142.html;jsessionid=PDkKJhsiQ5pJRjOKLzZRFnFB21d6Uw2eXrSIQfk4a3vxGT68Q7iw!1585849556.

Hazelwood, Nick. 2001. *Savage: The Life and Times of Jemmy Button.* New York: Thomas Dunne Books/St. Martin's.

Hebda, R. J., and R. W. Matthews. 1984. "Expansion of Western Red Cedar in Coastal Forests from 5000–2500 Years Ago Linked to Evolution of Woodworking Culture of PNW Tribes." *Science* 225 (4463): 711–13.

Henriksen, Edgar. 2014. "Norwegian Coastal Fisheries." Report 14/2014. Nofima, Tromsø, February. http://www.coastalfisheries.net/wp-content/uploads/2013/06/Norwegian-coastal-fisheries.pdf.

Hill, S. 2017. "'Codfather' Carlos Rafael Pleads Guilty." *National Fisherman*, March 30. https://www.nationalfisherman.com/northeast/codfather-pleads-guilty/.

Himes, A. H. 2010. "Small-Scale Sicilian Fisheries: Opinions of Artisanal Fishers and Sociocultural Effects in Two MPA Case Studies." *Coastal Management* 31 (4): 389–408.

Institute for Integrative Science and Health. 2012. "Elder Albert Marshall, HonDLitt—Mi'kmaw Nation." December 2. http://www.integrativescience.ca/uploads/files/Albert-Marshall-bioblurb2012-Two-Eyed-Seeing.pdf.

Ivanoff, Jacques. 2005. "Sea Gypsies of Myanmar." *National Geographic*, April. http://ngm.nationalgeographic.com/2005/04/sea-gypsies/ivanoff-text.

James, Jewell Praying Wolf. 2013. "The Search for Integrity in the Conflict over Cherry Point as a Coal Export Terminal." *Whatcom Watch*, August. http://www.whatcomwatch.org/pdf_content/LummiInsert.pdf.

Janes, Diane. 2003. "The Tulalip Salmon Ceremony." Special Collections, University of Washington Library.

Japan Times. 2013. "Aging 'Ama' Female Divers Strive to Revive Ranks." July 15. http://www.japantimes.co.jp/news/2013/07/15/national/aging-ama-female-divers-strive-to-revive-ranks/#.WXAwgyMrK2w.

Johannes, R. E. 1981. *Words of the Lagoon: Fishing and Marine Lore in the Palau District of Micronesia*. Berkeley: University of California Press.

King, A. E., and A. DeVogelaere. 2000. "Monterey Bay National Marine Sanctuary Final Kelp Management Report: Background, Environmental Setting, and Recommendations." October 3. https://montereybay.noaa.gov/research/techreports/trking2000.html.

Kirk, Ruth. 1986. *Tradition and Change in the Northwest Coast: The Makah, Nuu-chah-nulth, Southern Kwakiutl, and Nuxalk*. Seattle: University of Washington Press.

Klan, Yvonne Mearns. 1999. "The Apprenticeship of James Murray Yale." *British Columbia Historical News* 32 (4): 37–42.

Knutson, P. 1989. "The Unintended Consequences of the Boldt Decision." In *A Sea of Small Boats*, edited by John Cordell, 263–303. Cambridge, MA: Cultural Survival.

———. 2017. "Escaping the Corporate Net: Pragmatics of Small Boat Direct Marketing in the U.S. Salmon Fishing Industry of the Northeastern Pacific." *Marine Policy* 80:123–29. https://doi.org/10.1016/j.marpol.2016.03.015.

Korten, Fran. 2010. "Elinor Ostrom Wins Nobel for Common(s) Sense." *Yes! Magazine*, February 26. http://www.yesmagazine.org/issues/america-the-remix/elinor-ostrom-wins-nobel-for-common-s-sense?b_start:int=1&-C.

Koslow, J. A. 1982. "Limited Entry Policy and the Bristol Bay, Alaska Salmon Fishermen." *Canadian Journal of Fisheries and Aquatic Sciences* 39:415–25.

Kuoljok, Sunna, and John E. Utsi. 1993. *The Saami: People of the Sun and Wind*. Jokkmokk, Sweden: Ajtte, Swedish Mountain and Saami Museum.

Lätsch, Angelika. 2012. "Coastal Sami Revitalization and Rights Claims in Finnmark (North Norway): Two Aspects of One Issue? Preliminary Observations from the Field." *Senter for Samiske Studier—Skriftserie* 18:60–84.

Lawrence, Mike. 2016a. "Carlos Rafael, New Bedford's 'Codfather,' Indicted on 27 Counts." *Cape Cod Times*, May 9. http://www.capecodtimes.com/article/20160509/NEWS/160509439.

———. 2016b. "New Bedford Fisherman's Arrest Puts Spotlight on Industry's

Quota Question." *South Coast Today*, March 5. http://www.southcoasttoday.com/article/20160305/NEWS/160309618.

Lichatowich, Jim. 1999. *Salmon without Rivers: A History of the Pacific Salmon Crisis*. Washington, DC: Island Press.

———. 2013. *Salmon, People, and Place: A Biologist's Search for Salmon Recovery*. Corvallis: Oregon State University Press.

Loken, Linn Helene. 2014. "'Very High Inequality' in Chilean Society Breeds Resentment." *Santiago Times*, October 28. http://santiagotimes.cl/2014/10/28/very-high-inequality-in-chilean-society-breeds-resentment/.

Loki Fish Company. N.d. http://www.lokifish.com. Last accessed July 15, 2017.

Loomis, Ilima. 2015. "'Fishers' or 'Fishermen'—Which Is Right?" *Hakai Magazine*, October 13. https://www.hakaimagazine.com/article-short/fishers-or-fishermen-which-right?.

Lummi Island Wild. N.d. http://www.lummiislandwild.com. Last accessed July 15, 2017.

MacAulay, John M. 1998. *Seal-Folk and Ocean Paddlers: Sliochd nan Ròn*. Cambridge: White Horse Press.

MacEacheran, Mike. 2016. "The Last Mermaids of Japan." BBC, September 2. http://www.bbc.com/travel/story/20160829-the-last-mermaids-of-japan.

Macinko, Seth. 2014. "Lipstick and Catch Shares in the Western Pacific: Beyond Evangelism in Fisheries Policy?" *Marine Policy* 44:37–41.

MacLachlan, Moray, ed. 1998. *The Fort Langley Journals*. Vancouver: University of British Columbia Press.

Mapes, Linda. 2016. "Northwest Tribes Unite against Giant Coal, Oil Projects." *Seattle Times*, January 16. http://www.seattletimes.com/seattle-news/environment/northwest-tribes-unite-against-giant-coal-oil-projects/.

McDaniel, Nancy. 2004. "The Snohomish Tribe of Indians: Our Heritage, Our People." Special Collections, University of Washington Library.

McDonald, Danny. 2013. "Carlos Rafael and His Fish Are the American Dream." *Vice*, May 24. http://www.vice.com/read/carlos-rafael-fish-interview.

McGrath, David G., Alcilene Cardoso, Oriana T. Almeida, and Juarez Pezzuti. 2008. "Constructing a Policy and Institutional Framework for an Ecosystem-Based Approach to Managing the Lower Amazon Floodplain." *Environment, Development and Sustainability* 10 (5): 677–95. https://doi.org/10.1007/s10668-008-9154-3.

McKelvie, B. A. 1947. *Fort Langley: Outpost of an Empire*. Montreal: Southam Press.

McLean, Mark, manager, Regulatory Reviews, Fisheries Protection Program, to Steve Sanford, Environmental Assessment Branch, Nova Scotia Environment. 2016. "Proposed Environmental Effects Monitoring Program

2016-2020-Fundy Ocean Research Center for Energy (FORCE) and Cape Sharp Tidal Venture (CSTV)." June 14. Fisheries and Oceans, Canada, file 08-HMAR-MA7-00223. https://www.novascotia.ca/nse/ea/minas.passage.tidal.demonstration/DFO%20EEMP%20Recommendation%20Letter.pdf.

Mediterranean Platform of Artisanal Fishers. 2012. "Current Reform Will Cause Death of the Artisanal Fishing Sector, Says Mediterranean Platform of Artisanal Fishers." March 7. https://www.fishupdate.com/mediterranean-platform-of-artisanal-fishers-fishupdate-com/.

Mendenhall, Nancy Danielson. 2015. *Rough Waters: Our North Pacific Small Fishermen's Battle: A Fishing Family's Perspective*. Seattle, WA: Far Eastern Press.

Meriot, Christian. 1984. "The Saami Peoples from the Time of the Voyage of Ottar to Thomas van Westen." *Arctic* 37 (4): 373–84.

Miranda-Chumacero, G., R. Wallace, H. Calderon, G. Calderon, P. Willink, M. Guerrero, T. M. Siles, K. Lara, and D. Chuqui. 2012. "Distribution of Arapaima (*Arapaima gigas*) (Pisces: Arapaimatidae) in Bolivia: Implications in the Control and Management of a Non-native Population." *BioInvasions Records* 1 (1): 129–38.

Monterey Bay National Marine Sanctuary. 2014. "Resource Issues: Kelp Harvesting." March 5. http://montereybay.noaa.gov/resourcepro/resmanissues/kelp.html.

Montgomery, D. 2003. *King of Fish: The Thousand-Year Run of Salmon*. Boulder, CO: Westview Press.

Morton, Alexandra. 2016. "Opinion: Save Our Salmon—Get Diseased Fish Out of Pacific Ocean." *Vancouver Sun*, June 17. http://vancouversun.com/opinion/opinion-save-our-salmon-get-diseased-fish-out-of-the-pacific-ocean.

Moss, Madonna. 2011. *Northwest Coast: Archaeology as Deep History*. Washington, DC: Society for American Archaeology.

Mrozewski, Jo. 2015. Interview. edited by Kevin Bailey.

Muhlke, Christine. 2009. "Kelp Wanted." *New York Times*, August 2.

Murakami, Kery. 1995. "Initiative 640: A Bitter Fight over Dwindling Salmon Stocks." *Seattle Times*, September 6. http://community.seattletimes.nwsource.com/archive/?date=19951020&slug=2147941.

Myklebust, Anders. 2014. "Held gudsteneste mot Røkke-selskap" [Service held against Røkke's company]. *Vårt Land*, March 19. http://www.vl.no/2.616/held-gudsteneste-mot-rokke-selskap-1.16222.

National Oceanic and Atmospheric Administration. 2010. "NOAA Policy Encourages Catch Shares to End Overfishing and Rebuild Fisheries." November 4. http://www.noaanews.noaa.gov/stories2010/20101104_catchshare.html.

Naturvernforbundet. 2015. "Norwegian Seafood Industry and Environmentalists Slam Plans for Massive Pollution of Fjord." April 22. https://naturvernforbundet.no/pollution/norwegian-seafood-industry-and-environmentalists-slam-plans-for-massive-pollution-of-fjord-article33306-991.html.

Nelsen, Aaron. 2013a. "Chile's Fish Supply Decline 'Catastrophic' after Years of Overfishing." Pulitzer Center: *Global Post*, May 24. http://pulitzercenter.org/reporting/chile-pelluhue-riviera-marine-resources-fishing-law-drop-fish-populations-overfishing-quota-NFC.

———. 2013b. "Chile's Indie Fishermen Say New Law Favors Big Business." Pulitzer Center: *Global Post*, May 23. http://pulitzercenter.org/reporting/chile-quele-artisan-market-big-business-industrial-fishing-legal-advantage-corpesca-asipes.

Nelsen, Aaron, and Fernando Rodriguez. 2013. "Chile's Seafaring Oligopoly Threatens Artisan Fisheries." Pulitzer Center: *Global Post*, August 7. http://pulitzercenter.org/reporting/south-america-chile-south-pacific-fishing-fishermen-decimated-fish-populations-government-controversial-fisheries-law-jack-mackerel-hake-artisan-inequality.

Nickel-Kailing, Gail. 2015. "Loki Fish Company: Building a Brand, One Fish at a Time." *GoodFood World*, December 8. http://www.goodfoodworld.com/2015/12/loki-fish-company-building-a-brand-one-fish-at-a-time/.

North Pacific Fishery Management Council. 2016. "Twenty-Year Review of the Pacific Halibut and Sablefish Individual Fishing Quota Management Program." National Marine Fisheries Service. Final Draft, December. https://www.npfmc.org/wp-content/PDFdocuments/halibut/IFQProgramReview_417.pdf.

Northwest Treaty Tribes. 2013. "Lummi Nation Holds Reef Net Fishery at Cherry Point." September 17. http://nwtreatytribes.org/lummi-nation-holds-reef-net-fishery-cherry-point/.

———. 2014. "Lummi Fishermen Pass Down Reef Net Heritage." September 15. http://nwtreatytribes.org/lummi-fishermen-pass-reef-net-heritage/.

Norwegian Environmental Agency. 2014. "The Management Approach to Salmon Fisheries in Norway." Document no. CNL (14) 45. May. http://www.nasco.int/pdf/2014%20papers/CNL_14_45.pdf.

Odendahl, Terry. 2016. "Why Is Mine Waste Being Dumped Directly into the Ocean?" *Ecowatch: Insights*, March 4. http://www.ecowatch.com/why-is-mine-waste-being-dumped-directly-into-the-ocean-1882187670.html.

Odner, Knut. 1992. *The Varanger Saami: Habitation and Economy*. Oslo: Scandinavian University Press.

Opazo, Tania. 2016. "The Boys Who Got to Remake an Economy." Slate,

January 12. http://www.slate.com/articles/business/moneybox/2016/01/in_chicago_boys_the_story_of_chilean_economists_who_studied_in_america_and.html.

Opitz, Silvia, Julia Hoffmann, Martin Quaas, Nele Matz-Luck, Crispina Binohlan, and Rainer Froese. 2016. "Assessment of MSC-Certified Fish Stocks in the Northeast Atlantic." *Marine Policy* 71:10–14.

Orensanz, J. M., A. M. Parma, G. Jerez, N. Barahona, M. Montecinos, and I. Elias. 2006. "What Are the Key Elements for the Sustainability of 'S-Fisheries'? Insights from South America." *Bulletin of Marine Science* 76 (2): 527–56.

Osterblom, H., J. B. Jouffray, C. Folke, B. Crona, M. Troell, A. Merrie, and J. Rockstrom. 2015. "Transnational Corporations as 'Keystone Actors' in Marine Ecosystems." *PLOS ONE* 10 (5): e0127533. https://doi.org/10.1371/journal.pone.0127533.

Ostrom, Elinor. 1990. *Governing the Commons: The Evolution of Institutions for Collective Action*. New York: Cambridge University Press.

Otero, J., A. J. Jensen, J. H. L'Abee-Lund, N. C. Stenseth, G. O. Storvik, and L. A. Vollestad. 2011. "Quantifying the Ocean, Freshwater and Human Effects on Year-to-Year Variability of One-Sea-Winter Atlantic Salmon Angled in Multiple Norwegian Rivers." *PLOS ONE* 6 (8): e24005. https://doi.org/10.1371/journal.pone.0024005.

Pauly, Daniel J. 1999. "Fisheries Management: Putting Our Future in Places." In *Fishing Places, Fishing People*, edited by Dianne Newell and Rosemary E. Ommer, 355–62. Toronto: University of Toronto Press.

Pedersen, Steinar. 2012. "The Coastal Sami of Norway and Their Rights to Traditional Marine Livelihood." *Arctic Review on Law and Politics* 3 (1): 51–80.

Pembroke, Timothy. 1979. "An Ethnohistorical Report Regarding the Usual and Accustomed Fishing Grounds of the Tulalip Tribes." Report to the Tulalip Tribes.

Pesca Sicilia. N.d. "Reti a circuizione per lampughe" [Fishing gear for lampuga]. http://www.pescasicilia.net/index_239.html. Last accessed July 15, 2017.

Pinkerton, Evelyn, and Reade Davis. 2015. "Neoliberalism and the Politics of Enclosure in North American Small-Scale Fisheries." *Marine Policy* 61: 303–12. https://doi.org/10.1016/j.marpol.2015.03.025.

Pinkerton, Evelyn, Kim Olsen, Joy Thorkelson, Henry Clifton, and Art Davidson. 2016. "You Thought We Canadians Controlled Our Fisheries? Think Again." TheTyee.ca, January 11. https://thetyee.ca/Opinion/2016/01/11/Who-Controls-Fisheries/.

Pollan, Michael. 2008. *In Defense of Food: An Eater's Manifesto*. New York: Penguin.

Port Orford Sustainable Seafood. N.d. "Our Story." https://www.posustainable seafood.com/caught-in-oregon/our-story/. Last accessed July 15, 2017.

Raines, Ben. 2016. "Kingpins of the Gulf Make Millions off Red Snapper Harvest without Ever Going Fishing." AL.com, January 24. http://www.al.com/news/index.ssf/2016/01/kingpins_of_the_gulf_make_mill.html.

Real Good Fish. N.d. http://www.realgoodfish.com/. Last accessed July 15, 2017.

Real News Network. 2013. "The Privatization of Chile's Seas." January 18. http://therealnews.com/t2/index.php?option=com_content&task=view&id=767&Itemid=74&jumival=9528.

Riley, Jonathan. 2016. "Cost of Tidal: Bay of Fundy Fishermen Worried about Fish Stocks." *Digby Courier*, May 17. http://www.novanewsnow.com/News/Local/2016-05-17/article-4532130/Cost-of-tidal%3A-Bay-of-Fundy-fishermen-worried-about-fish-stocks/1.

Roberson, Leslie. 2015. "Fishing on Ghana Man Time." *Newfound* 6 (1). https://newfound.org/archives/volume-6/issue-1/nonfiction-leslie-roberson/.

Roberts, Callum. 2012. *The Ocean of Life: The Fate of Man and the Sea*. New York: Viking Press.

Roman, Joe. 2013. "Fishing with Dolphins." *Slate*, January 31. http://www.slate.com/articles/health_and_science/science/2013/01/fishing_with_dolphins_symbiosis_between_humans_and_marine_mammals_to_catch.html.

Saegrov, H., K. Hindar, S. Kaias, and H. Lura. 1997. "Escaped Farmed Atlantic Salmon Replace the Original Salmon Stock in the River Vosso, Western Norway." *ICES Journal of Marine Science* 54:1166–72.

Sang-Hun, Choe. 2014. "Hardy Divers in Korea Strait, 'Sea Women' Are Dwindling." *New York Times*, March 29. http://www.nytimes.com/2014/03/30/world/asia/hardy-divers-in-korea-strait-sea-women-are-dwindling.html.

Scheer, Roddy. 2007. "Passion Fish." *Seattle*, September, 178–85.

Schlosser, Thomas P. 1978. "Washington's Resistance to Treaty Indian Commercial Fishing: The Need for Judicial Apportionment." October. http://www.msaj.com/papers/commfish.htm.

Schwartz, Ralph. 2015a. "Coal Port Would Lift Whatcom Economy." *Bellingham Herald*, June 22. http://www.bellinghamherald.com/news/local/article25204258.html.

———. 2015b. "For Lummis, Fishing Is More Than a Living—It's a Way of Life." *Bellingham Herald*, November 14. http://www.bellinghamherald.com/news/local/article44514417.html.

Seafood News. 2015. "Brazil Wants to Increase Paiche Production to Raise Farmed Fish Output to 2 Million Tons by 2020." April 2. http://www.seafoodnews.com/Story/970368/Brazil-Wants-to-Increase-Paiche-Production-to-Raise-Farmed-Fish-Output-to-2-Million-Tons-by-2020.

Seattle Times staff. 2013. "American Seafoods Fined over Alleged Deception." *Seattle Times*, June 1, Business/Technology. http://www.seattletimes.com/business/american-seafoods-fined-over-alleged-deception/.

Shields, Captain Ed. 2001. *The Salt of the Sea: The Pacific Coast Cod Fishery and the Last Days of Sail*. Lopez Island, WA: Pacific Heritage Press.

Shoreline Journal. 2016. "We Don't Think So." October. http://www.theshorelinejournal.com/oct1623.pdf.

Skogvang, Susann Funderud. 2014. "Extractive Industries in the North—What about Environmental Law and Indigenous Peoples Rights." *Nordic Environmental Law Journal* 2014 (1): 13–19.

Smith, Carsten. 2014. "Fisheries in Coastal Sami Areas: Geopolitical Concerns?" *Arctic Review on Law and Politics* 5 (1): 4–10.

Smith, Julian. 2014. "Infinite Depths." *World Wildlife Magazine*, Fall. https://www.worldwildlife.org/magazine/issues/fall-2014/articles/infinite-depths.

Southern Pacific Review. 2015. "Fishing Law in Chile." October 6. http://southernpacificreview.com/2015/10/07/fishing-law-in-chile/.

———. 2016. "Chile's Corrupt Fishing Law." June 11. http://southernpacificreview.com/2016/06/11/chiles-corrupt-fishing-law/.

Spicuzza, Mary. 1998. "Sea No Evil." *Metro Santa Cruz*, June 11–17. http://www.metroactive.com/papers/cruz/06.11.98/kelp-harvesting-9823.html.

Star Online. 2014. "Malaysia's Last 'Fish Listeners.'" August 20. http://www.thestar.com.my/news/nation/2014/08/20/malaysia-fish-listeners/.

Stein, Julie. 2000. *Exploring Coastal Salish Prehistory*. Seattle: University of Washington Press.

Stewart, Hilary. 1977. *Indian Fishing: Early Methods on the Northwest Coast*. Seattle: University of Washington Press.

St. Fleur, Nicholas. 2014. "Sustainable Abalone: Monterey Company Wins Plaudits from Environmentalists." *San Jose Mercury*, March 25.

Stoll, Steven. 2016. "No Man's Land." *Orion*, January/February. https://orionmagazine.org/article/no-mans-land/.

Subsecretaría de Pesca y Acuicultura. 2016. "Estado de situación de las principales pesquerías chilenas, año 2015" [State of the situation of the principal fishes of Chile]. Edited by División de Administración Pesquera Departamento de Pesquerías.

Survival International. N.d. "The Ocean Is Our Universe." http://www.survivalinternational.org/galleries/moken-sea-gypsies. Last accessed July 15, 2017.

Suttles, Wayne. 1954. "Post-Contact Culture Change among the Lummi Indians." *British Columbia Historical Quarterly* 18 (2): 29–102.

Suzuki, David. 2014. "Salmon Farms: Has Anything Changed after a Decade of Controversy?" David Suzuki Foundation, May 1. http://www.davidsuzuki

.org/blogs/science-matters/2014/05/salmon-farms-has-anything-changed-after-a-decade-of-controversy/.

Swan, James. (1857) 1972. *The Northwest Coast*. Seattle: University of Washington Press. Original edition, Harper & Brothers.

Taranger, G. L., O. Karlsen, R. J. Bannister, K. A. Glover, V. Husa, E. Karlsbakk, B. O. Kvamme, K. K. Boxaspen, P. A. Bjorn, B. Finstad, A. S. Madhun, H. C. Morton, and T. Svasand. 2014. "Risk Assessment of the Environmental Impact of Norwegian Atlantic Salmon Farming." *ICES Journal of Marine Science* 72 (3): 997–1021. https://doi.org/10.1093/icesjms/fsu132.

Tarver, Toni. 2015. "Sea-ing a Better Way to Feed the World." *Food Technology Magazine*, August. http://www.ift.org/Food-Technology/Past-Issues/2015/August.aspx.

Thomas, Tim. 2014. *The Abalone King of Monterey*. Mt. Pleasant, SC: History Press.

Thorstad, Eva B., and Torbjørn Forseth, eds. 2015. "Status for norske laksebestander i 2015" [Status of salmon populations in Norway in 2015]. Report no. 8. Vitenskapelig råd for lakseforvaltning [Scientific Advisory Committee for Atlantic Salmon Management], Trondheim, June. http://www.nina.no/archive/nina/PppBasePdf/Rapp%20Vitr%C3%A5dlaks/Thorstad%20Status%20RVitr%C3%A5d%20laks2015-8.pdf.

Thrush, Coll-Peter. N.d. "The Lushootseed Peoples of Puget Sound Country." Digital Collections, Topical Essays, Special Collections, University of Washington Library. http://content.lib.washington.edu/aipnw/thrush.html. Last accessed July 30, 2017.

Tizon, Alex. 1999. "The Boldt Decision/25 Years: The Fish Tale That Changed History." *Seattle Times*, February 7. http://community.seattletimes.nwsource.com/archive/?date=19990207&slug=2943039.

Tsikliras, A., A. Dinouli, V. Z. Tsiros, and E. Tsalkou. 2015. "The Mediterranean and Black Sea Fisheries at Risk from Overexploitation." *PLOS ONE* 10 (3): e0121188. https://doi.org/10.1371/journal.pone.0121188.

Undercurrent News. 2014. "Rokke Defends 'Nostalgic' Loss-Making Fishing Investments." April 10. https://www.undercurrentnews.com/2014/04/10/rokke-defends-nostalgic-loss-making-fishing-investments/.

United Nations. 2008. "United Nations Declaration on the Rights of Indigenous Peoples." Report no. UN07-58681. March. http://www.un.org/esa/socdev/unpfii/documents/DRIPS_en.pdf.

United Nations Economic and Social Council. 2010. "Report on Indigenous Fishing Rights in the Seas with Case Studies from Australia and Norway." January 8. http://www.un.org/esa/socdev/unpfii/documents/E.C.19.2010.2EN.pdf.

United Nations Food and Agriculture Organization. 2012. "The State of World Fisheries and Aquaculture." Fisheries and Aquaculture Department. http://www.fao.org/docrep/016/i2727e/i2727e.pdf.

———. 2013. "Technical Consultation on International Guidelines for Securing Sustainable Small-Scale Fisheries." May 20–24. Document no. TC-SSF/2013/2. ftp://ftp.fao.org/FI/DOCUMENT/ssf/SSF_guidelines/TC/2013/2e.pdf.

———. 2014. "State of World Fisheries and Aquaculture: Opportunities and Challenges." http://www.fao.org/3/a-i3720e.pdf.

———. N.d. "Technical Assistance and Cooperation: GFCM Framework Programme (FWP)." General Fisheries Commission for the Mediterranean (GFCM). http://www.fao.org/gfcm/activities/technical-assistance-and-cooperation/fwp-info/en/. Last accessed April 23, 2016.

United Nations Regional Information Centre for Western Europe. 2012. "'Ocean-Grabbing' as Serious a Threat as 'Land-Grabbing': UN Expert on Right to Food." October 31. http://unric.org/en/latest-un-buzz/28003-ocean-grabbing.

Van der Voo, Lee. 2014. "The Big Fish Wins Again." *Slate*, May 2. http://www.slate.com/articles/health_and_science/science/2014/05/catch_shares_investment_firms_are_taking_over_the_fishing_rights_system.html.

———. 2016. *The Fish Market: Inside the Big-Money Battle for the Ocean and Your Dinner Plate*. New York: St. Martin's Press.

Vasilakopoulos, P., C. D. Maravelias, and G. Tserpes. 2014. "The Alarming Decline of Mediterranean Fish Stocks." *Current Biology* 24 (14): 1643–48. https://doi.org/10.1016/j.cub.2014.05.070.

Veta la Palma. N.d. "Veta la Palma: Parque Naturale." http://www.vetalapalma.es/index.asp?LG=2. Last accessed July 15, 2017.

Vollset, Knut Wiik, Bjorn Barlaup, Shad Mahlum, Bjornar Skar, Helge Skoglund, Eirik Normann, Jens Christian Holst, Vidar Wennevik, and Georg Skaet. 2014. "Migration and Predation of Atlantic Salmon Smolts from Vosso." Uni Research, Bergen, Norway.

Walton, Izaak. 1653. *The Compleat Angler*. London.

Washington Historical Quarterly. 1934. "The Indian Chief Kitsap." 25 (4): 297–301.

Weddell, James. 1825. *Voyage towards the South Pole, Performed in the Years 1822–'24*. London: Longman, Hurst, Rees, Orme, Brown, and Green.

Westerdahl, Christer, ed. 2010. *A Circumpolar Reappraisal: The Legacy of Gutorm Gjessing*. Oxford: Archaeopress. http://www.academia.edu/1461936/CircumSami.

Wickens, Jim. 2016. "How Vital Fish Stocks in Africa Are Being Stolen from Human Mouths to Feed Pigs and Chickens on Western Factory Farms."

Independent, September 17. http://www.independent.co.uk/news/world/africa/how-vital-fish-stocks-in-africa-are-being-stolen-from-human-mouths-to-feed-pigs-and-chickens-on-a7234636.html.

Wilder, R. J. 1998. *Listening to the Sea: The Politics of Improving Environmental Protection*. Pittsburg: University of Pittsburgh Press.

Wilkinson, Charles. 2000. *Messages from Frank's Landing: A Story of Salmon, Treaties, and the Indian Way*. Seattle: University of Washington Press.

Wilson, B., R. S. Batty, F. Daunt, and C. Carter. 2006. "Strategic Environmental Assessment of Marine Renewable Energy Development in Scotland: Collision Risks between Renewable Energy Devices and Mammals, Fish and Diving Birds: Report to the Scottish Executive." Scottish Association for Marine Science, November 25. http://nora.nerc.ac.uk/504110/1/N504110CR.pdf.

Windh, Jacqueline. 2016. "The Man Who Became the Rainbow." *Hakai Magazine*, March 11. https://www.hakaimagazine.com/article-short/man-who-became-rainbow.

Wong, Venessa. 2014. "Why Whole Foods Wants You to Eat This Giant Amazonian Fish." Bloomberg, August 28. http://www.bloomberg.com/news/articles/2014-08-28/whole-foods-is-pushing-paiche-in-order-to-support-fish-farms.

Woods, Fronda. 2005. "Who's in Charge of Fishing?" *Oregon Historical Quarterly* 106 (3): 412–41.

World Fishing & Aquaculture. 2013. "A New Kind of Future." August 3. http://www.worldfishing.net/news101/regional-focus/a-new-kind-of-future.

World Wildlife Fund. 2001. "The Status of Wild Atlantic Salmon: A River by River Assessment." May. http://d2ouvy59p0dg6k.cloudfront.net/downloads/salmon2.pdf.

Wright, Tony. 2016. "Opinion: Fundy Needs a Test Turbine to Answer Valid Questions." *Chronicle Herald*, June 8. http://thechronicleherald.ca/opinion/1370881-opinion-fundy-needs-a-test-turbine-to-answer-valid-questions.

Wu, Olivia. 2007. "Abalone's Luster Grows." *San Francisco Chronicle*, February 28.

Zambito, Thomas. 2014. "Was That Salmon Born in Chile or Scotland? Buyer Says She Was Pressured to Mislabel Fish." *Star-Ledger*, May 4.

Zappes, C. A., A. Andriolo, P. C. Simoes-Lopes, and A. P. Madeira di Benditto. 2011. "Human-Dolphin (*Tursiops truncatus* Monagu, 1821) Cooperative Fishery and Its Influence on Cast Net Fishing Activities in Barra de Imbe/Tramandai, Southern Brazil." *Ocean & Coastal Management* 54:427–32.

Index

Aker Seafoods, 105
Alaska, Gulf of, 112, 156
Alaska pollock, 5, 8, 197
Alaska Sustainable Fisheries Trust, 205
Allende, Salvador, 51, 53
ama divers, 68
Amazone (company), 183–89
American Abalone Farms, 220n2 (chap. 9)
American Fisheries Act (1998), 105
American Seafoods, 105, 197
Annapolis Royal, 137, 138
anti-salmonera, 58
aquaculture: Chile, 58; growth, 13–14; importance, 5, 13
Arana, Patricio, 56–58
Arantes, Caroline, 179–83, 186, 189f
arapaima, 189f; alternative names, culture, 183–89; fishery, 179, 181–83; life history, 179–81; natural history, 178–79; restocking, 185–86
artisanal fishery, definition, 4
ArtisanFish, 187
Atlantic salmon: Chile, 58, 67; collapse, 35, 41–44; farming, 46–47, 90–91. See also *Salmo salar*

Barents Sea, 93, 95, 108
Barlaup, Bjorn, 38–39, 45

Bay of Fundy Inshore Fishermen's Association, 141
Bering Expedition, 163
Bivdi, 94, 109–10; definition, 217n2
Bojer, Johan, 111
Boldt, George Hugo, 87–88
botos-bons, 190
Broughton Archipelago, 174
Button, Jemmy (O'run-del'lico), 213–14n6
Buzzards Bay, 174

Calise, Tonino, 1–3
Cannery Row, 169, 203
cannizzi, 32f, 33
Cape Sharp Tidal, 133–34, 137–38, 142
Caputo, Vincenzo, 26–27
Cárdenas, Juan Carlos, 49–56, 67
Carlsbad Aquafarm, 191
Catalina Sea Ranch, 191
catch shares: advantages, 11–12; disadvantages, 11–12; history, 9–12; other programs, 198–99; quota determination, 196; successful programs, 197–98
Cherry Point, WA, 125, 126
Chicago Boys, 53, 55
Chile: aquaculture, 58; fisheries status, 55; history, 50–51

Chiloé, 52, 58, 213n4
Chinook (salmon), 75, 84, 126, 147, 149, 153, 154, 202
chum salmon: fishing method, 149–50; life history, 147–49; management, 154–55; status, 149
Clean Ocean Action, 142
Coast Salish: and colonists, 82–84; and Europeans, 80–82; prehistory, 78–80; relationship with salmon, 78–80
Cobb, Leesa, 199
Common Fisheries Policy, 29
community-based fisheries, 199–201
Community Fisheries Network, 203
consolidation of fisheries, 29, 195, 199
Cook, James, 80
Cormorant Island, 174
Coryphaena hippurus, 26. *See also* lampuga
Cuttyhunk Shellfish Farms, 174–75

Dadswell, Mike, 131, 136, 137, 138
da Gama, Vasco, 9
Darwin, Charles, 64–65, 213n6
Department of the Environment, Nova Scotia, 135, 137
Donation Land Claim Acts of 1850 and 1853, 82
Dorry, Niaz, 194–96, 205
Dover, Harriette Shelton, *Tulalip, from My Heart*, 88, 216n10, 217n21
Duncan, David James, *River Teeth*, 142
Dutch East India Company, 10

Earth Ocean Farms, 191
Ecology Action Centre, 135
Eco Océanos, 49–50
Ecotrust, 203
Eidfjord, Norway, 37
Elliott Bay, WA, 145, 146f

Emera, 134, 141
Endangered Species Act, 126
Environmental Defense Fund (EDF), 11, 197, 200
Esselen, 163
Eythórsson, Einar, 105

Fay, Trevor, 162f, 167–68, 170–72
Finnmark Act of 2005, 105
Finnmark Coastal Fisheries Commission, 106
First Salmon ceremony, 71–74, 88–89
fisheries: expansion, 7; history, 6–8; indigenous, 7; jobs, 5; value, 5
Fisheries Law of 2012 (Chile), 52
fish listeners, 128
FitzRoy, Robert, 213–14n6
Food and Agriculture Organization (FAO), 8, 13, 143
Fort Langley, 72–73, 81, 164, 215n4, 217n20
Frank, Billy, Jr., 72, 87, 88
Fraser River, Canada, 72, 75, 79, 81, 119, 125, 217n20
Friedman, Milton, 53, 66
Fryberg, Ray, 89, 215n6, 216n9, 216n10
Fundy, Bay of, 130f; biodiversity, 131–32; energy, 133–34; flow, 129; legends, 139, 143; lobsters, 141; turbines, 136, 140
Fundy Ocean Research Center for Energy (FORCE), 133–40, 141; environmental effects monitoring plan (EEMP), 135; partners, 133; role, 133
Furnes, Helge, 35, 38–41, 40f, 45–46

Garfield, Dorothy, 174
gaspereau, 133
Gherson, Isaac, 183, 185, 187
Greenpeace, 195, 221n1 (chap. 11)

Grotius, Hugo, *Mare Liberum*, 10
Gusinde, Martin, 65

haenyeo divers, 67
Haida, 75
Halifax, Nova Scotia, 136
Hanuse, Debra, 90
Hines, Maria, 154
Huchoosedah, 85
Hudson's Bay Company, 72, 81, 215n4

IFQ (individual fisheries quota), 10–12; Chile, 50, 55, 61, 63; Gulf of Mexico, 197–98; halibut, 198. *See also* ITQ (individual transferable quota)
industrial fisheries, 4–8; Chile, 54–55; growth, 25; Norway, 101, 103
Industrial Revolution, 4
International Collective in Support of Fishworkers, 13
International Labor Organization's Convention No. 169 of 1989, 109
International Pacific Halibut Commission, 157
International Union for Conservation of Nature (IUCN), 182
Ischia: fishery, 2; history, 20–22
Italy, traditional fisheries, 5, 29, 33
ITQ (individual transferable quota), 11, 29. *See also* IFQ (individual fisheries quota)

James Beard Foundation, 127
Johannes, Robert, *Words of the Lagoon*, 93

keystone species, definition, 161
Kirk, Ruth, 85
Kirouac, Ian, 117, 121, 123
Knutson, Pete, 145–46, 147f, 149–52, 155

Kuterra, 90–91
Kvalsund, Norway, 93, 103
Kwakiutl, 85, 217n20
Kwakwaka'wakw, 75, 85, 90

Lacco Ameno, Italy, 2, 19, 21, 26, 30
Laguna, Brazil, 190
Laich-Kwil-Tach, 75, 85–86
lampuga, 26–27, 33. See also *Coryphaena hippurus*
Laplander (Lapp), 100–101
"Law of the Sea" (UN), 10
Leaman, Bruce, 157
Legoe Bay, WA, 117, 118f, 122, 124–25
Ley Longueira, 49–52, 57–58, 61
Lofoten Islands, 96, 107, 111
Loki Fish Company, 146, 153
Longueira, Pablo, 51–52
Lovewell, Alan, 201–2
Lummi Island, 117, 120, 124, 126–27
Lummi tribe, 28, 80, 82, 85; activism, 125–26; fishery, 125
Lushootseed, 83

Magallanes, Chile, 67
Magnuson–Stevens Act (MSA), 203
Maine, Gulf of, 129, 132
Maitencillo, Chile, 60–62
Makah tribe, 80, 85
Malay, 127
Mapuche, 50, 52, 213
Mare Liberum (Grotius), 10
Matthews, Steve, 151
McNeil, Stephen, 141
Mediterranean Platform of Artisanal Fishers (MedArtNet), 205
Mi'kmaq, 139, 142–43, 215n3
Minas Basin, 129, 130f, 131–32
Minas Passage, 129, 130f, 135, 137–38, 140, 142
Moken, 127–28
Monterey Abalone Company, 162, 162f, 167–72

Monterey Bay Fisheries Trust, 203
Monterey Bay National Marine Sanctuary (MBNMS), 169–70
Monterey Fishing Community Sustainability Plan, 203
Moss Landing Marine Laboratories, 168, 171

'Namgis, 90; clam gardens, 174
Naples, Gulf of (also Bay of Naples), 20, 20f, 206
National Council for the Defense of the Fishing Heritage of Chile (CONDEPP), 205
National Marine Fisheries Service (NMFS), 11, 143
Neruda, Pablo, 66–67
Nimpkish River, 90, 174
Nisqually tribe, 72, 85, 87
Northern Straits Salish, 119, 215n7
"North of Falcon," 154
North Pacific Fishery Management Council, 156–57, 221n1 (chap. 11)
Northwest Atlantic Marine Alliance (NAMA), 194–95, 199
Norwegian Fishermen's Association, 106, 109
Norwegian Institute of Marine Research, 104
Norwegianization, 99, 110
Nova Scotia Department of Energy, 133–34, 141
Nova Scotia Power, 133–34, 137, 141
Nussir, 103

Offshore Energy Research Association of Nova Scotia, 133, 141
Olsen, Torulf, 94–95, 98, 100–101, 103–4, 110
Organisation for Economic Co-operation and Development (OECD), 66
Ostrom, Elinor, 204–5

Pacific cod, 112–13
Pacific Fishery Management Council, 201
Pacific halibut: Bering Sea bycatch conflict, 157; IFQ, 198–99
Pacific salmon: colonization of the Pacific Northwest, 75–78; decreasing population, 75–78
paiche, 177, 179, 185, 187, 189. See also arapaima
Palau, 193–94, 204
pesca d'ombra, 32f, 33
Pinochet, Augusto, 51–54, 60
Point Elliott, Treaty of, 82–85, 216n14
Point Roberts, WA, 120–21, 124
Porter, Darren, 129–33, 130f, 136, 140–43
Port Orford, OR, 199–201
Port Orford Ocean Resource Team, 199
Port Orford Sustainable Seafood, 201
Pozzuoli, Italy, 2
Procida, Italy, 2, 24–25
Puget Sound: archaeology, 79; biodiversity, 132; chum salmon, 115, 145, 147–49, 154–55; and Coast Salish, 69, 72; geology, 77; salmon fishing, 84, 86, 88, 117, 126

Qaba'xad, 73–75, 85–86, 215n5

Rafael, Carlos, 196–97, 222n4
Real Good Fish (RGF), 201
red abalone: culture, 166–67, 171–72; El Niño, 172; fishery, 165–66; interaction with otters, 165–66; kelp wars, 169–70; withering foot syndrome, 165
reefnet: history, 119–21, 124–25; operation, 121–24
Regno di Nettuno, 23

River Teeth (Duncan), 142
Roberson, Leslie, 143–44
Røkke, Kjell Inge, 105, 218n15

Salish Sea, 71, 115, 117, 125
Salmo salar, 46. *See also* Atlantic salmon
Sámi: fight for fishing rights, 104–6, 109–10; loss of fishing rights, 93, 101–3; modern history, 98–100; and Norwegian fishery reforms, 103, 107; pre-history, 95–98; and red king crab, 107–9
Sámi Act of 1987, 100
Scientific Advisory Committee for Atlantic Salmon Management, 43
Seafood Watch, 117, 173–74, 195, 202
sea gypsies, 127–28
sea otters, history of exploitation, 163–65
Seavey, Art, 162f, 167–70, 172
sitjenot, 35, 36f, 41
Skagit tribe, 75, 85
Skykomish tribe, 73, 80, 83, 85
Slow Fish movement, 206, 211n2
slow food movement, 13, 27, 30
small-scale fishery, definition, 4
Snohomish tribe, 71, 73–75, 80–81, 83, 85–86, 215n4, 215nn6–7, 216n10
Snoqualmie tribe, 80, 83
SOS Marine Conservation Foundation, 90
Sproul, Colin, 141–42
Stevens, Isaac Ingalls, 82–85, 88, 216n13
Stevens, John Paul, 88
Stillaguamish tribe, 83
Sundstrom, John, 127

teflubenzuron, 47
Tesoura, 190

Thimble Island Ocean Farm, 191
Tierra del Fuego, 63, 65–66, 67, 206, 213–14n6
Tlingit, 75
traditional fishery: definition, 4; importance of, 3–4; problems, 3
tramaglio, 23, 24
treaties: and fishing rights, 86–88; Point No Point, Quinault, Medicine Creek, Neah Bay, 83. *See also* Point Elliot, Treaty of
Troncoso, Miguel, 59–62, 59f
Tulalip, from My Heart (Dover), 88, 216n10, 217n21
Tulalip tribes, 71, 82–83, 86–89
TURF (territorial-use rights fisheries), 60–61, 204
Turner, Matthew, 112

UNESCO World Heritage site, 95
Uni Research, 38, 212n2
United Nations Declaration on the Rights of Indigenous Peoples of 2007, 109

Valparaíso, Chile, 56, 59–60
Vespoli, Paolo, 21–23, 25–26, 28–31, 31f
Veta la Palma, 191
Vollset, Knut, 38–39, 45–46
Voss, Norway, 38
Vosso River, 38f; collapse of salmon, 35, 43–44

Washington Department of Fish and Wildlife, 117, 149, 151
West African fisheries, 143–44
Wetzel, Blaine, 127
Whole Foods (Market), 177, 187, 189, 221n6
Words of the Lagoon (Johannes), 193–94
World Bank, 29

World Forum of Fisher Peoples, 205
WWF (World Wildlife Fund), 186

Yale, James Murray, 72–75, 81, 215nn4–5
Yámana, 206, 213nn4–5, 214n6

Yoshiyuki, Iwase, 68
Yup'ik, 147–48
Yurimaguas, Peru, 178f, 183

Zapallar, Chile, 56
Zarate, Ricardo, 187, 221n7